Springer Earth System Sciences

Series Editors

More information about this series at http://www.springer.com/series/10178

Sara F. Oldfield • Peggy Olwell
Nancy Shaw • Kayri Havens

Seeds of Restoration Success

Wild Lands and Plant Diversity in the U.S.

Springer

Sara F. Oldfield
Cambridge, UK

Nancy Shaw
USDA Forest Service
Rocky Mountain Research Station
Boise, ID, USA

Peggy Olwell
Bureau of Land Management
Washington, DC, USA

Kayri Havens
Chicago Botanic Garden
Glencoe, IL, USA

ISSN 2197-9596 ISSN 2197-960X (electronic)
Springer Earth System Sciences
ISBN 978-3-319-96973-2 ISBN 978-3-319-96974-9 (eBook)
https://doi.org/10.1007/978-3-319-96974-9

Library of Congress Control Number: 2018951066

This Springer imprint is published by the registered company Springer Nature Switzerland AG
The registered company address is: Gewerbestrasse 11, 6330 Cham, Switzerland

Foreword

My first direct experience with the essential role of native plants in helping ecosystems recover occurred 25 years ago, as I took a late-spring walk in a Montana old-growth forest in which a hundred-thousand-acre, stand-replacing wildfire had burned for weeks the previous summer into early autumn. The fire had only gone out when snow began to fall in mid-September at the valley level, as is common in that extreme landscape. The following spring, I wanted to see if anything would be growing from the ashes. The site was one with which I was very familiar, one that I had loved and known well as a naturalist, for the rare wild orchids that grew beneath the towering canopy of hemlocks and Western red cedars. It had been one of my favorite places to take my young daughters and learn about plants together. That forest lay on public land managed for conservation. It had never been logged; the fire allowed to burn as part of a federal policy of allowing natural processes like wildfire to occur on such lands.

I was not yet an ecologist back then, as I walked around that burned forest in awe. I thought I was going there to mourn what once was. What I found instead was a wild riot of native plants growing from the bare black, mineral soil in the places the fire had burned with the highest severity. This astonishing array was dominated by species such as fireweed, spirea, beargrass, and Scouler's willow—which years later I discovered are colonizers that help rebuild and stabilize soil, so that other plant species can come in later and thrive.

Over the years, I returned to that site with my mentor, US Forest Service fire ecologist Steve Wirt, who had a passion for native plants, to observe the native plant community succession—huckleberries coming in—and more. In due time, that experience and others like it shaped my ecological sensibility. And over the years, that site and the native plants it contained inspired me to attend graduate school and study how everything in a food web, from native plants to apex predators such as wolves, is connected.

Flash forward 15 years, to a private Society for Ecological Restoration and Wildlands Network symposium on the High Lonesome Ranch in Colorado. Embedded in Bureau of Land Management (BLM) lands, this working ranch managed for conservation lies at the core of some of the most degraded public and private lands in the USA. As the ranch research director, I studied how drought combined with elk movement patterns and cattle grazing were affecting the grassland. Unsurprisingly, I had found that the ranch rangelands consisted of 95% non-native grass species. At the symposium, I had the good fortune to meet Peggy Olwell, one of the co-authors of *Seeds of Restoration Success*. She didn't have to do much to convince me that native plant restoration was the key to restoration success here and elsewhere.

Flash forward another 5 years, and I had completed my graduate education and was working as the Chief Scientist at Earthwatch Institute, where I directed a global research program. In that position, I discovered that the lessons I'd learned about the importance of native plants in northwest Montana and Colorado were universal and applied equally whether I was in Mongolia, or Kenya, or the Scottish Highlands.

Quite simply, plants are foundational to the existence of all other life on Earth and to the wellbeing of humanity, because of their intrinsic role in the carbon cycle. Native plants in particular serve a key evolutionary role in sustaining diverse, healthy ecosystems and in providing the ecosystem services (e.g., food, water, and energy security) on which humans rely for survival. Yet their importance is often overlooked. In *Seeds of Restoration Success,* the authors, eminent botanists who among them have many decades of experience leading native plant conservation in the USA and globally, argue compellingly why managing, conserving, and restoring the landscapes, ecosystems, and vegetation types of the USA requires a new awareness of the plant diversity on which people depend.

The authors of this book have worked steadily and tirelessly within their respective botanic gardens and government agencies to raise awareness of the importance of native plants to the resilience of ecosystems worldwide. To that end, they have founded the US National Seed Strategy and Seeds of Success Program (the US national native seed collection program), led plant science and conservation at the Chicago Botanic Garden, led a botany research program at the U.S. Forest Service and at the Royal Botanic Gardens, Kew, and lead global conservation initiatives for the IUCN and other organizations. Their lifework comes from the heart, and this book is a manifestation of that. Yet the authors also embody the highest level of rigor as empirically trained scientists who have boldly and innovatively advanced native plant science. Over the years, I have taken in their lessons in my work and learned new lessons on my own that have only served to emphasize the importance of native plant conservation and native seeds.

As a restoration ecologist who primarily works in fire-prone wild lands in North America, I have had many powerful lessons about how after a seemingly catastrophic event, such as a large, severe wildfire, native plants show up in the form of soil seed banks or wind-borne seeds, or root sprout, and help heal the Earth. Without colonizers such as reedgrass and lamb's quarters and other natives, we'd be facing decades of degradation and erosion. And without wild lands that have a vigorous population of native flora, which enables them to respond with resilience to pervasive environmental stressors such as drought, and a restoration strategy for native plants and their seeds on those lands, ecological integrity and functionality across a spectrum of ownership would be compromised.

More than seven decades ago, famed US ecologist and conservationist Aldo Leopold wrote, "The last word in ignorance is the man who says of an animal or plant, 'What good is it?' If the land mechanism as a whole is good, then every part is good, whether we understand it or not. If the biota, in the course of eons, has built something we like but do not understand, then who but a fool would discard seemingly useless parts? To keep every cog and wheel is the first precaution of intelligent tinkering" (Leopold 1986, 190). Such an approach is essential to restore degraded ecosystems and manage them for resilience.

The world is changing rapidly, ecosystems becoming increasingly compromised by climate change and a burgeoning human population. We face many challenges to restoring and maintaining native plant communities, but have made huge progress even in the last decade. Broad in scope geographically and in terms of biotic communities, *Seeds of Restoration Success* provides an invaluable tool and bright beacon of hope as we reimagine and restore US wild lands to create a more resilient future. May the deep wisdom this book contains help us sustain native plants in US landscapes and beyond.

Cristina Eisenberg
Concord, MA, USA
September 14, 2019

Reference

Leopold A (1986) A Sand County almanac: and sketches here and there. Ballantine Books, New York

Preface

The idea for this book grew from our discussions on the need for an ambitious agenda for promoting the importance of plant diversity and the need to use native plant diversity wisely. All four authors have been privileged to enjoy long careers in plant conservation and ecological restoration whether in the federal government sector, botanic garden community, or with non-governmental organizations. We all enjoy plants in their natural habitats and also enjoy the practical aspects of growing plants in our gardens! Our collective hope has been for management of native plant diversity to become mainstream – is recognized as equally important as the management of any other natural resource, such as wildlife, water, energy, or minerals. This hope has not yet become reality, but there have been major steps forward in the twenty-first century in international policy, legislation, scientific understanding, and practical biodiversity management. We therefore remain optimistic.

The scope of this book is ambitious, and we have, by necessity, been quite selective in the information included. We have attempted to provide an overview of the knowledge of plant diversity, vegetation types, and ecosystems across the United States. People form an integral part of ecosystems, and we have covered the threats to, and management and restoration of native plant diversity in a summarized overview. The science behind plant conservation and ecological restoration has not been covered in depth, but the bibliographic references given should help in providing additional supporting information. In researching the book, we have drawn on rich resources of botanical and ecological publications, many of which are available online. We are deeply impressed by the often unsung efforts of botanists and ecologists working across the United States to research, conserve, and restore the national plant heritage. We hope this book will help to explain and honor their invaluable contributions as we move into an uncertain future.

Cambridge, UK — Sara F. Oldfield
Washington, DC, USA — Peggy Olwell
Boise, ID, USA — Nancy Shaw
Glencoe, IL, USA — Kayri Havens

Acknowledgments

We are grateful to all our botanical colleagues in botanic gardens, universities, non-government organizations, private sector, and government for the inspiration behind this book and for so generously contributing their expertise and information. We truly appreciate Dr. Cristina Eisenberg for her careful review of an earlier draft of the book. The provision of beautiful images is gratefully acknowledged. We thank Tom Kaye and Carolyn Menke for preparing the map for Fig. 7.1. We thank Emily Yates for preparing the maps and Botanic Gardens Conservation International for information used in Fig. 4.2. It has been a pleasure to work together on this project and we acknowledge that any errors in the publication are our own. The views and opinions contained in this book belong to the authors and do not represent official views of our agencies or institutions.

Contents

1 **Introduction** 1
References 5

2 **The Importance of Native Plants and the Pressures They Face** 7
Plants for Ecosystem Services 9
Plants for Food 10
Plants for Medicine 11
Plants for Ornamental and Landscape Use 13
Plants for the Economy 15
Plants Under Threat 16
Habitat Loss 17
Invasive Species 17
Overexploitation 18
Climate Change 19
Where are the Plant Diversity Hotspots? 20
References 22

3 **The Importance of Ecosystems** 25
Forests 27
Woodlands 32
Grasslands 33
Deserts 34
Wetlands 36
Changing Ecosystems 38
References 38

4 **Conservation of Plant Species** 41
Conservation Assessment, Planning, and Action 42
In Situ Conservation 44
Ex Situ Conservation 50
Collaborative Action 56
References 56

5 **Land Ownership, Land Management, and Land Use** ... 59
Land Ownership ... 59
Federal Land Management ... 61
Management of Non-federal Land ... 64
Management of Land for Agriculture ... 66
Management of Land for Forestry ... 71
Management of Land for Recreation and Conservation ... 73
Urbanization ... 75
References ... 75

6 **Restoring the Future** ... 77
Producing Plants for Ecological Restoration ... 80
Collection and Storage of Seed ... 81
Agricultural Seed Production ... 82
Selection of Native Plant Species for Restoration ... 84
Development of Seed Zones ... 86
Native Plants for Restoration of Forests ... 87
Native Plants for Restoration of Prairies ... 89
Native Plants for Restoration of Deserts ... 90
Native Plants for Restoration of Coastal Wetlands ... 94
National Action: Supporting Pollinators ... 96
National Action: The National Seed Strategy ... 98
References ... 100

7 **Looking Ahead** ... 103
Progress in Ecological Restoration ... 104
The Need for Seed ... 106
Overcoming Challenges ... 109
Overcoming Plant Blindness ... 110
The Green Future ... 113
References ... 113

Appendix 1: Federal Environmental Laws and Policy ... 115

Appendix 2: Internationally Important Wetland Sites of the U.S. Listed on the Ramsar Convention ... 125

Index ... 129

Scientific Index ... 135

Acronyms

AOSCA	Association of Official Seed Certifying Agencies
APGA	American Public Gardens Association
ARRI	Appalachian Regional Reforestation Initiative
ARS WRPIS	USDA Agricultural Research Service Western Regional Plant Introduction Station
BGCI	Botanic Gardens Conservation International
BLM	DOI Bureau of Land Management
BSE	USDA Forest Service, Bend Seed Extractory
CBG	Chicago Botanic Garden
CEQ	Council on Environmental Quality
CERCLA	Comprehensive Environmental Response, Compensation and Liability Act
CFLR	Collaborative Forest Landscape Restoration
CMPMC	USDA NRCS Cape May Plant Materials Center
CNPLX	California Native Plant Link Exchange
CPC	Center for Plant Conservation
CPNPP	DOI BLM Colorado Plateau Native Plant Program
CRP	USDA NRCS Crop Reserve Program
CWR	Crop wild relatives
DOI	Department of the Interior
EPCA	Energy Policy and Conservation Act
ESA	Endangered Species Act
FANN	Association of Florida Native Nurseries
FAO	Food and Agriculture Organization of the United Nations
FWCA	Fish and Wildlife Coordination Act
FLPMA	Federal Land Policy and Management Act
FWS	DOI U.S. Fish and Wildlife Service
G rank	Global rank – a conservation status ranking system
GAP	USGS Gap Analysis Program
GEA	Grasslands Ecological Area
GOSR	New York Governor's Office of Storm Recovery

GSPC	UN Global Strategy for Plant Conservation
HUD	U.S. Department of Housing and Urban Development
IUCN	International Union for the Conservation of Nature
MARSB	Mid-Atlantic Regional Seed Bank
NAAQS	National Ambient Air Quality Standards
NAS	National Academy of Sciences
NCBG	North Carolina Botanical Garden
NCC	Nanocrystalline cellulose
NCGRP	USDA ARS National Center for Genetic Resources Preservation
NEPA	National Environmental Policy Act
NEWFS	New England Wild Flower Society (now Native Plant Trust)
NFMA	National Forest Management Act
NFWPCAS	National Fish, Wildlife, and Plants Climate Adaptation Strategy
NGO	Non-governmental organization
NHPA	National Historic Preservation Act
NOAA	Department of Commerce National Oceanic and Atmospheric Administration
NPGS	USDA ARS National Plant Germplasm System
NPS	DOI National Park Service
NRCS	USDA Natural Resources Conservation Service
NTBG	National Tropical Botanical Garden
NWFP	Northwest Forest Plan
NWI	National Wetlands Inventory
PCA	Plant Conservation Alliance
PDR	Purchase of development rights program
PEP	Hawaii Plant Extinction Prevention Program
PMC	USDA NRCS Plant Materials Center
RCA	Soil and Water Resources Conservation Act
RDA	Rural Development Act
RNA	Research Natural Area
SARA	Superfund Amendments and Reauthorization Act
SCS	USDA Soil Conservation Service (now the USDA Natural Resources Conservation Service)
SDGs	UN Sustainable Development Goals
SER	Society for Ecological Restoration
SFBE	San Francisco Bay Estuary
SOS	Seeds of Success Program
SST	Seedlot Selection Tool
STRIPS	Iowa State University's Science-based Trials of Rowcrops Integrated with Prairie Strips Project
TgC	Trillion grams of carbon
UNEP WCMC	United Nations Environmental Programme, World Conservation Monitoring Centre
USBG	U.S. Botanic Garden
USDA	U.S. Department of Agriculture

US DOT FHWA	U.S. Department of Transportation, Federal Highway Administration
USFS	USDA Forest Service
USGCRP	U.S. Global Change Research Program
USGS	DOI U.S. Geological Survey
WHIP	USDA NRCS Wildlife Habitat Incentives Program
WRI	World Resources Institute
WWETAC	USDA Forest Service, Western Wildland Environmental Threat Assessment Center

Chapter 1
Introduction

> *The human world is a long way from meeting the needs of the present, and it is borrowing massively from the future – not only by piling up money debt, but also by degrading the resources from which all real wealth ultimately comes.* (Donella Meadows 1992)

The United States (U.S.) has an extraordinary range of landscapes, ecosystems, and vegetation types, varying from desert to rainforest and from Arctic tundra in northern Alaska to subtropical swamps in southern Florida. The land has been modified by humans to some extent since prehistoric times, but there is still a magnificence of natural diversity to manage and celebrate. The very destiny of the U.S. has been linked with its spectacular landscapes nature symbolizing national greatness from the time of American independence. Geologic and climatic forces have sculpted the landscapes traversed today. Plants provide a surface covering, define vegetation and ecosystems, and give a visual identity to localities across the nation. Plants are woven into the very fabric of human existence, but their fundamental importance is frequently overlooked. In this book we discuss how and why managing, conserving, and restoring the landscapes, ecosystems, and vegetation types of the U.S. requires a new awareness of the plant diversity on which people depend.

Managing land sustainably requires a deep understanding of ecological factors across time and space. Critically, this knowledge is still limited. What we do know, however, is that the native plant species that help to define ecosystems and vegetation types provide ecosystem services that sustain people and their economies. Native plant communities are key to ecosystem integrity and resilience. Natural diversity, the wide array of different plant species and populations within different communities, enhances resilience and adaptability to environmental change. As major and fundamental elements of biodiversity, native plant species have intrinsic and irreplaceable biotic value that will become increasingly important in the future. Plants are able to capture the sun's energy and create oxygen from carbon dioxide in the process of photosynthesis, powering all life on the planet. They are critical in the sequestration of carbon, to the water cycle, and to controlling climate.

S. F. Oldfield et al., *Seeds of Restoration Success*, Springer Earth System Sciences, https://doi.org/10.1007/978-3-319-96974-9_1

As discussed in the next chapter, plant species also provide essential resources for food, medicines, and a host of industrial products.

Flowering plants and conifers, whether trees, shrubs, grasses, or forbs, produce seed as a means to perpetuate the existence of their species. Seeds contain an embryonic plant, along with the nutrients it needs to develop into a seedling, maintained in a state that allows it to wait until the environment is favorable for germination. Seeds are a precious resource enabling the maintenance and recovery of natural ecosystems. They enable adaptation to changing environmental conditions because of the genetic variability they contain. Conserving the diversity of native plants and their seeds is essential to ensure that ecological options are available in the future. The different approaches to plant conservation are outlined in Chap. 4. While seed research is still needed, we demonstrate how seed banking or storage is an efficient and cost-effective component in the complex management of biodiversity.

The need to conserve ecosystems, species, and genetic diversity as part of land management has been recognized in the U.S. for well over 100 years. Food production, timber harvesting, the extraction of energy and minerals, industrialization, and urbanization all impact the natural environment and lead to the biodiversity conservation imperative. Globally, the main threats to biodiversity are habitat loss, invasive species, over-exploitation, pollution, and climate change. Now with the impacts of global change being increasingly accepted, the urgency for conservation action has become more apparent, and efforts must be made to further develop coordinated responses. Climate change is having a major impact on the distributions and interactions of plant and animal species worldwide. Climate change is a factor leading to local, national, and global extinctions.

The changing global climate is leading to increasingly intense weather events including hurricanes, flooding, drought, and wildfires, which affect plant and animal species, ecosystems, and human welfare. Ecological destruction, fatalities, and loss of livelihoods are interlinked consequences. From 1980 to 2012, disaster-related losses amounted to \$3.8 trillion worldwide. Nearly 90 percent of these reported disasters, with 74 percent of financial costs (\$2.8 trillion) and 61 percent of lives lost (1.4 million in total) were caused by extreme weather (World Bank 2013). In the U.S., there have been 219 weather and climate disasters since 1980 with costs of \$1 billion or more and total costs exceeding \$1.5 trillion (Smith 2018). There have been significant changes in extreme weather events over the last several decades. Heavy downpours have increased nationally, leading to increased flood events. Heat waves have become more frequent and intense, especially in the West. Droughts in the Southwest and heat waves across the country are projected to become more intense (Walsh et al. 2014).

Connected with increased drought, the number of large wildfires in the U.S. has nearly doubled since the 1980s, and the average length of the wildfire season has grown significantly. Wildfire risk depends on factors such as temperature, soil moisture, and the condition of the vegetation. All these factors are linked directly or indirectly to climate change. The acreage of land burned annually in the U.S. has also increased significantly since the 1980s. According to National Interagency Fire Center data, nine of the ten years with the largest areas burned have occurred since 2000, including the peak year of 2015 when over 10 million acres (4 million ha) burned (U.S. Environment Protection Agency 2016).

Individual large fires are generally catastrophic for habitats, native plant and animal species, local livelihoods, and local health. In 2015, the Soda Fire burned nearly 280,000 acres (110,000 ha) in southwest Idaho and southeast Oregon, including nearly 200,000 acres (81,000 ha) of sage-grouse (*Centrocercus urophasianus*) habitat, portions of 41 grazing allotments, three wild horse management areas, and a popular recreation area. Due to the location of the fire and its effect on federal, state, and private lands, a collaborative approach was taken to assess values at risk within the entire burned area (BLM 2006). In the immediate aftermath of the fire, assessment focused on mitigating threats to life, property, and resources within the burned area over the following 3 years.

The intensity, frequency, and duration of North Atlantic hurricanes have all increased since the early 1980s. The relative contributions of human and natural causes to these increases remain uncertain, but hurricane-associated storm intensity and rainfall rates are projected to increase as the climate continues to warm (Walsh et al. 2014).

Extreme weather events, in turn, affect biodiversity and the ecological resilience that it provides. The importance of native plant species and plant communities with their intimate connection to location, is not always taken into account in land use and conservation planning. In this book, we aim to stress why native plants and their seeds are so important as part of land management and for addressing climate change. Actions to conserve and restore native plant diversity, including having appropriate seed resources readily available, are important strategic components in adapting to climate change and being prepared for emergencies related to extreme weather events.

Globally, the need to maintain biodiversity and restore ecosystems as a response to climate change is recognized in United Nations Biodiversity Targets agreed upon in Japan in 2010 and known as the Aichi Targets (Secretariat of the Convention on Biological Diversity 2010). Aichi Target 15 of the UN Strategic Plan for Biodiversity states that:

> By 2020, ecosystem resilience and the contribution of biodiversity to carbon stocks has been enhanced, through conservation and restoration, including restoration of at least 15 per cent of degraded ecosystems, thereby contributing to climate change mitigation and adaptation and to combating desertification.

The Global Strategy for Plant Conservation, which directly supports the UN Strategic Plan for Biodiversity, gives specific targets for plant species and plant-rich habitats (Convention on Biological Diversity 2012).

In the U.S., legislation and policies to manage and restore ecosystems to increase ecological carbon sequestration and reduce overall levels of greenhouse gas emissions have been developed, but their future is currently less certain. The Energy Independence and Security Act of 2007 remains the relevant legislation (see Appendix 1). The use of native plant species in ecosystem restoration and for adaptation and mitigation is emphasized in this Act. At a national level, policies for adapting to climate change and for emergency preparedness to cope with more extreme weather events–hurricanes, flooding, drought, and wildfires – were introduced. In the U.S. more than 30 federal agencies developed Climate Change Adaptation Plans.

Building on these plans, Executive Order 13653 Preparing the United States for the Impacts of Climate Change (see Appendix 1) was signed in 2013 directing federal agencies to

- Modernize federal programs to support climate-resilient investments
- Manage lands and waters for climate preparedness and resilience
- Provide information, data, and tools for climate change preparedness and resilience
- Plan for climate change related risk

Executive Order 13653 recognized natural resources as critical for the economy, health, and quality of life. It directed agencies, *"to identify changes that must be made to policies, programs, and regulations to strengthen the climate resilience of watersheds, natural resources, and ecosystems, and the communities and economies that depend on them."* Federal agencies were required to protect and better promote natural storm barriers, such as dunes and wetlands, as well as to protect the carbon sequestration benefits of forests and other lands to help reduce the carbon pollution that contributes to climate change. Despite the significant progress in linking healthy ecosystems with climate resilience, Executive Order 13653 was rescinded in 2017.

Native plant communities that provide ecological resilience continue to be severely impacted by habitat modification, including the spread of invasive plant species and altered wildfire regimes, which are influenced by the changing climate. The increasingly frequent extreme weather events have major implications for land management. The National Academy of Sciences (NAS) defines resilience as *"the ability to prepare and plan for, absorb, recover from, and more successfully adapt to adverse events"* (Linkov et al. 2014). Genetically diverse plant communities hold the key to adaptation. Securing plant resources helps prepare and plan for future adverse weather events and allows options for adaptive land management. Seeds, whether stored naturally in soil seed banks or dried and stored in specially constructed seed banking facilities, provide the raw materials for restoration. Central to this book is the importance of seeds as a natural resource. Before we consider the management of land, plant species, and seed resources in the U.S., we summarize the importance and appreciation of plant species in Chap. 2 and the diversity of ecosystems that seeds help to build and maintain in Chap. 3.

Although U.S. ecosystems and the species they contain have been degraded in a relatively short period of time as outlined in Chap. 5, we emphasize that there are many positive actions to restore and manage land sustainably as discussed in Chap. 6. Native plant species are increasingly appreciated for their beauty, utility, and economic potential. Looking ahead, the diversity of native plants needs to be recognized as a significant natural resource, and the skills and expertise to manage this resource need to be enhanced.

References

Bureau of Land Management (BLM) (2006) Soda Fire emergency stabilization and rehabilitation, Idaho and Oregon. https://www.blm.gov/sites/blm.gov/files/Program_FishandWildlife_WildllifeIdahoSodaFireESR_StatusReport.pdf. Accessed 3 May 2018

Convention on Biological Diversity (2012) Global strategy for plant conservation: 2011–2020. Botanic Gardens Conservation International, Richmond,UK

Linkov I, Bridges T, Creutzig F, Decker J, Fox-Lent C, Kroger W, Lambert JH, Levermann A, Montreuil B, Nathwani J, Nyer R, Renn O, Scharte B, Scheffler A, Scheurs M, Thiel-Clemen T (2014) Changing the resilience paradigm. Nature Clim Change 4(6):407–409

Meadows, D. (1992) Donella Meadows Archive. http://donellameadows.org/archives/what-does-sustainability-mean/. Accessed 25 February 2019

Secretariat of the Convention on Biological Diversity (2010) Strategic plan for biodiversity 2011–2020 and the Aichi Targets. https://www.cbd.int/doc/strategic-plan/2011-2020/Aichi-Targets-EN.pdf. Accessed 24 April 2018

Smith AB (2018) 2017 U.S. billion-dollar weather and climate disasters: a historic year in context. NOAA National Centers for Environmental Information. https://www.climate.gov/news-features/blogs/beyond-data/2017-us-billion-dollarweather-and-climate-disasters-historic-year. Accessed 3 April 2018

U.S. Environmental Protection Agency (2016) Climate change indicators in the United States. Fourth Edition. EPA 430-R-16-004. www.epa.gov.climate-indicators. Accessed 3 April 2018

Walsh J, Wuebbles D, Hayhoe K, Kossin J, Kunkel K, Stephens G, Thorne P, Vose R, Wehner M, Willis J, Anderson D, Doney S, Feely R, Hennon P, Kharin V, Knutson T, Landerer F, Lenton T, Kennedy J, Somerville R (2014) Ch. 2: our changing climate. In: Melillo JM, Richmond TC, Yohe GW (eds) Climate change impacts in the United States: the third national climate assessment. U.S. Global Change Research Program, pp 19–67. https://doi.org/10.7930/J0KW5CXT

World Bank (2013) Building resilience: integrating climate and disaster risk into development. Lessons from World Bank Group experience. The World Bank, Washington, DC

Chapter 2
The Importance of Native Plants and the Pressures They Face

From food to medicine, from clean air to clear water, to intrinsic beauty and the very landscape that covers our planet, plants define the human experience. But despite this indispensable and multifaceted role in our lives, plants are remarkably undervalued. (John Clark 2017)

Plants are important for our survival, our well-being, and our inspiration. The ecosystems and native flora of the U.S. are resources of immense importance, underpinning all life in the country. Plants provide a vast array of ecosystem goods and services. They regulate water supplies, provide atmospheric oxygen, and sequester carbon, controlling the atmosphere and climate of our planet. In addition, they provide food for humans, habitat for wildlife, food for pollinators that are essential for crop production, timber and raw materials for industry, and energy production are just a few of the fundamental benefits provided by plants.

Forests with majestic trees and desert landscapes with towering cacti inspire a sense of wonder, and some of the most beautiful wildflower sites in the world are found in the U.S. Mount Rainier National Park in Washington, for example, provides spectacular alpine and sub-alpine meadows and tundra brimming with wildflowers. It has been described by Gibbons (2011) as possibly the most flowery place in the world! Individual wild plant species, such as the magnificent redwoods and saguaro cacti, have iconic status and signify a strong sense of place. Plants contribute hugely to our enjoyment of the natural world, and more fundamentally, can act as indicators of the health of the planet. Degraded habitats, with low native plant diversity, impair ecosystem services and result in an impoverished future. Action is needed now to keep our options open.

Flowering plants, whether grasses, forbs, shrubs, or trees produce seed – a critical natural resource and an asset that deserves greater recognition in light of the ecological challenges we face. Native seed is the main vehicle for migration of plant species to areas favorable for species survival, and it also enables adaptation of species to changing environmental conditions. Another feature of the seed of many species is the capacity to persist in the soil, contributing to the maintenance of ecosystem stability. Capturing and storing this seed resource, either by carefully tending natural areas or

S. F. Oldfield et al., *Seeds of Restoration Success*, Springer Earth System Sciences, https://doi.org/10.1007/978-3-319-96974-9_2

by other means, has been a good land stewardship practice. Now, as we increasingly understand the economic and ecological values of plants and the future options they provide, it is ever more important to manage native seed resources wisely.

The native plant resource of the U.S. consists of over 18,000 species with new plants still being discovered and described. At a state level, California has the highest diversity of plant species, followed by Texas, Arizona, New Mexico, and Oregon. During the 1980s, between five and ten new plants to science were described from the California flora each year (Shevock and Taylor 1988).

Scientific documentation of the native U.S. flora began in the seventeenth century with accounts by visiting European botanists, such as Robert Morison, Leonard Plukenet, and John Ray. These early plant explorers were eager to discover new horticultural varieties for ornamental gardens. Fredrick Pursh's *Flora Americae Septentrionalis* (1814) was the first broad account of the flora on a continental scale. Botanists living in the U.S. began to document native plants in the early 1800s, and production of a comprehensive flora was attempted for the first time by John Torrey and Asa Gray in the 1830s. The U.S. National Herbarium was established at the Smithsonian Institution in 1848 when the first plant collections were accessioned from a U.S. expedition of the Pacific coast and surrounding lands. Around 50,000 specimens of 10,000 species were collected. Throughout the nineteenth century, botanists worked to produce floras for different regions of the U.S. In 1905 N. L. Britton of the New York Botanical Garden began publication of the monographic series of plant families for the combined *North American Flora*.

In the 1960s, leading botanists decided to work together on a comprehensive flora of North America. With initial funding from the National Science Foundation, the *Flora of North America*, collaborative project was established at the Smithsonian Institution. This led to the establishment of the current *Flora of North America* project at the Missouri Botanical Garden in 1983. This major ongoing project draws on the expertise of plant taxonomists in North America and elsewhere, contributing their knowledge. Ultimately, *Flora of North America* will provide a comprehensive account of the diversity of plant species in the U.S. with species descriptions and maps. At the same time, various regional flora projects are underway or completed, covering California, the Intermountain West, the Pacific Northwest, the Great Plains, Texas, and the Southeast. Baseline information on plant taxonomy and species distribution is essential to fully understand and describe U.S. botanical resources.

Box 2.1: Crested Butte Wildflower Festival

The mountains around Crested Butte, Colorado, a former mining town, form a magnificent area for alpine wildflowers. A State Senate Resolution passed in 1990 designated this area in the Central Colorado Rockies as the "Wildflower Capital of Colorado." Each year Crested Butte holds a Wildflower Festival to celebrate the stunning displays of aspen sunflowers (*Helianthus quinquenervis*), lupines (*Lupinus* spp.), paintbrushes (*Castilleja* spp.), gentians (*Gentiana* spp.), penstemons (*Penstemon* spp.) and blue

(continued)

Box 2.1 (continued)
columbine (*Aquilegia coerulea*), the state flower. These flowers form part of the open grassland vegetation, generally at higher altitudes than the coniferous forests and below the tundra belt. The mission of the Crested Butte Wildflower Festival is to advocate for the preservation and appreciation of wildflowers in the Gunnison Valley and adjacent wilderness areas through exploration, education, and identification of wildflowers and their role in a unique environment. Thousands of people take part in the Festival every year

Sources: Gibbons (2011), CBWF (2019).

Plants for Ecosystem Services

Native plant diversity is essential for the provision of ecosystem services, including the regulation of carbon, energy, and water cycles. Native plant species help to buffer the impacts of climate change and are essential for developing future responses. Unfortunately, the plant diversity we depend on at ecosystem, species, and genetic levels is under threat, and as a result, the foundations of our natural world are crumbling. The major threats to plants in the U.S. are described below. Actions to conserve and restore native plant diversity, including having appropriate seed resources readily available, as described more fully in Chaps. 4, 5, and 6, are important strategic components of climate change adaptation and emergency preparedness. They have an impact on national and global security.

The uptake of carbon dioxide by green plants during photosynthesis is the major pathway by which atmospheric carbon is captured and stored. As a consequence, plants and their assemblages have a significant role in influencing climate. The carbon storage of different ecosystems is described in Chap. 3.

Native plants have a major role in controlling water supplies. As noted by Hatfield et al. (2014), evapotranspiration from soil, plants, and water bodies, together with transpiration, the use and release of water from plants, is the second largest component of the water cycle after precipitation. Evapotranspiration responds to temperature, solar energy, winds, humidity in the atmosphere, and moisture availability at the land surface. In turn, evapotranspiration regulates amounts of soil moisture, groundwater recharge, and runoff. Transpiration by plants comprises between 80 percent and 90 percent of total evapotranspiration on land. Resilient plant-diverse ecosystems are important to capture, control, and filter water, helping to enable stability of water supply.

Maintaining watershed vegetation is a key strategy in regulating the flow of water to towns and cities. Approximately 9 million people in New York City and adjacent areas benefit from access to clean, inexpensive drinking water. About 90 percent of that water is drawn from the Catskill/Delaware watershed where the abundant forest reserves, as well as soil with adequate carbon levels, provide excellent conditions for natural water filtration (WRI 2008).

One of the most widely recognized ecosystem services provided by native plant species is their role in supporting pollinators. The attributed value of crops that are directly dependent on insect pollination in the U.S. has been estimated at $15 billion (Calderone 2012). Honey bees (*Apis mellifera*), the most recognizable pollinators, were introduced by early European settlers in the 1620s. They have been in serious decline for more than three decades. There are over 4,000 other bee species, both native and introduced, together with a wide range of other insect, bird, and mammal pollinators. Pollinator populations depend directly on plant populations, especially native plants.

Plants for Food

Provisioning services are another category of ecosystem services where biodiversity provides goods of direct value to humans. Plant species are, for example, essential for the global food supply. Around 5,000 plant species continue to supply nutritional needs of people around the world. In the U.S., Native Americans used a wide range of wild and cultivated plants in their diets. Over 1,600 plant species used as food by Native Americans have been recorded. Essential food plants included corn (*Zea mays*), squash (*Cucurbita* spp.) and beans (*Phaseolus* spp.). Others have included blueberries and cranberries (*Vaccinum* spp.), hazelnuts (*Corylus* spp.), acorns from native oak (*Quercus* spp.) trees (a staple of Californian Indians), pinyon seeds (*Pinus* spp.), wild rice (*Zizania* spp.) and other grains, cacti, and oil producing plants such as sunflowers (*Helianthus* spp.).

The common sunflower (*Helianthus annuus*) is a plant of major economic and cultural importance. Native Americans began cultivating the sunflower thousands of years ago. Selection over time has produced the domesticated sunflower with single large seed heads and enlarged oil-rich seeds. The sunflower is now the fourth most important oilseed crop in the world and is grown in every temperate region. In all, 52 species of wild sunflowers are native to North America, with most of these species endemic to the U.S. Several species are now very rare in the wild. The wild sunflower species are adapted to a wide range of habitats and have much greater variability for many traits than the cultivated sunflower. Modern sunflower crop breeding programs have used wild species as the source of a number of traits, including resistance to downy mildew, rust, sunflower moth, and other pests and diseases.

The estimated economic contribution of the wild species to the cultivated sunflower is between $269 and $384 million per year. Continued use of the tremendous genetic diversity present in wild sunflower populations is critical to meeting the future challenges of emerging pests and environmental changes that may threaten sunflower production (USDA 2014). The Threatened Pecos sunflower (*Helianthus paradoxus*) of New Mexico and Texas is, for example, considered an important source of salinity tolerance.

As with wild species of sunflower, wild relatives of other crop species are generally of vital importance for improving food and other cultivated crops. A national inventory of crop wild relatives (CWR) of the U.S. has been produced listing 4,600 wild plant taxa useful for breeding as well as wild species directly used for food, forage, medicine, herbs, ornamentals, or for environmental restoration (Khoury et al. 2013). The list includes some non-native CWR that have become naturalized in the U.S. From this list a high priority selection of 285 taxa from 30 genera was developed reflecting wild species very closely related to major food crops - including relatives of sunflower, walnut (*Juglans* spp.), squash, wild rice, and plum (*Prunus* spp.) (Greene et al. 2015). In an era of rapid global change, drought, salt tolerant, and disease resistant genes in wild plants will be essential to prepare agricultural crops for the future.

Wild plant communities still contribute directly to our daily food through both products gathered from the wild and the provision of genetic material for breeding. Furthermore, they provide the supporting framework for agriculture through the ecosystem services they provide. As noted above, native vegetation is important to maintain water supplies and also to prevent soil erosion in agricultural regions.

Box 2.2: Texas Wild Rice

Texas wild rice (*Zizania texana*) is an aquatic perennial grass that only grows in the clear waters of the San Marcos River in Texas. This species is a wild relative of the more widespread and commercially produced wild rice species *Zizania aquatica* and *Zizania palustris*. These two species were historically harvested by Native Americans, some of whom still consider the grain sacred or of ceremonial value. Today, wild rice grain – primarily *Zizania palustris* – is grown commercially in the U.S., Australia, and Hungary. In Canada, large stands are harvested for food from lake margins, especially in Manitoba.

Texas wild rice is a potentially invaluable resource of hardy genetic material. The Texan species was so abundant in the 1930s that it was considered a pest by the local irrigation company as it impeded waterflow in drainage ditches. Now this species is listed as Endangered at the state and federal levels, and only a few populations remain in the wild. Decline has mainly been caused by pumping water out of the Edwards Aquifer for agriculture, industry, and domestic use. This has directly affected the spring flow of the San Marcos River. Fortunately, both in situ and ex situ conservation measures have been designed for this important species.

Sources: Richards et al. (2007), CBD (2019).

Plants for Medicine

Even as healthcare and pharmaceutical developments become ever more sophisticated, wild plants remain of huge importance as a source of medicine. In the U.S. almost 3,000 different plants have been used traditionally for their medicinal properties. American ginseng (*Panax quinquefolius*) was used by Native Americans to treat fevers, coughs, and headaches. Black cohosh (*Cimicifuga racemosa*), a native plant of eastern woodlands and a staple of Cherokee medicine, served many purposes, ranging from use as a diuretic to a cure for rheumatic pains. Bloodroot (*Sanguinaria canadensis*) provided the Cherokee with a red dye and a medicine to cure coughs and inflammation of the lungs. Blue cohosh (*Caulophyllum thalictroides*), another eastern woodland plant, helped cure toothaches for the Cherokee, while the Chippewa used the root to treat cramps. Goldenseal (*Hydrastis canadensis*) has been collected as a medicinal plant for centuries. Traditionally, the plant was used by Native Americans as a yellow dye and as a treatment for a variety of ailments ranging from arrow wounds to cancer. Among North American tribes, goldenseal is now mainly used by the Cherokee and Iroquois. All these traditionally used plants, together with many more, remain important for healthcare and the rural economy today.

Currently at least 175 plants native to North America are for sale in non-prescription medicinal markets in the U.S. Overall in the U.S., the annual market for medicinal herbs is worth more than $3 billion, and the demand is growing. More than 60 million consumers in the U.S. take herbal remedies. More doctors are recommending herbal medicines and some health insurance plans offer coverage for herbal treatments. Many of the medicinal plant products are collected from the wild in large quantities because cultivation techniques have not been developed or are not considered economical. For some species, such as black cohosh, goldenseal, and American ginseng, there are well-established export markets. With careful management to prevent over-harvesting, wild collection of native species can provide an important source of income in rural areas.

European settlers in the eighteenth century used a root wash made from goldenseal to treat eye inflammation, and this species remained a common ingredient of patent medicines into the late nineteenth century. With the introduction of antibiotics, U.S. domestic use of goldenseal declined, but it continued to be a valuable commodity in Europe. Over the past 20 years, with resumed interest in herbal medicines, goldenseal has regained popularity in the U.S. It is now one of the most popular medicinal plants used in North America and remains a staple of most herbal practitioners. Current uses include treatment for a wide range of disorders. Goldenseal products available in trade include cut and powdered root, tincture, fluid extract, powdered extract, tea, and a wide variety of processed pharmaceutical remedies in the form of capsules, tablets, salves, and other products. It is used in a range of multi-ingredient products. Leaves and stems also have commercial value. Most of the material produced continues to be sourced from the wild, but cultivation is expanding with propagation either by rhizome division or from seed.

American ginseng, a native plant species of North American eastern woodlands, is still highly sought after for its medicinal value. Each year this long-lived perennial herb produces a green shoot from a short underground rhizome, which grows from a fleshy taproot. Ginseng harvesters remove the root, which kills the plant. Commercial harvesting of wild ginseng plants has occurred since the 1700s, mainly for export to China. In 2011, about 22,300 kg of wild ginseng were exported from the U.S. FWS (2013).

The current value of exports of American ginseng is approximately $27 million annually (FWS 2015).

The contribution of American ginseng to rural economies is significant, although its estimated value is only about 2 percent of hardwood timber value. It has been estimated that the annual sale of American ginseng could support a thousand full-time harvesters, providing them with a mean annual income of $25,000. Integrating collection of American ginseng into forest management has the potential to enhance livelihoods of rural communities while also protecting biodiversity (Chamberlain et al. 2013).

The purple coneflower (*Echinacea purpurea*) is a herbaceous perennial that is native to the prairies and open woodlands of eastern and central U.S. Purple coneflower was commonly used by Native Americans for hundreds of years before the arrival of European explorers and colonizers. By the early 1800s, purple coneflower became a popular herbal remedy among U.S. settlers, and soon became commonly used in Europe as well. Its popularity increased significantly after research was carried out in Germany in the 1920s. Now extracts from this species are among the most commonly used herbal remedies to strengthen the immune system against colds and flu in both the U.S. and Europe. The marketed plant parts include roots, leaves, and flowers, with roots generally the most sought after. Collecting from the wild has put pressure on natural populations. Other species of coneflower used medicinally, include narrow-leaf coneflower (*Echinacea angustifolia*) and yellow coneflower (*Echinacea paradoxa*). Roots and seed of yellow coneflower have been harvested from wild populations in Arkansas and Missouri.

Cultivation of coneflowers as a commercial crop has been developing over the past 20 years with the Pacific Northwest the main area of production (Sari et al. 1999).

Black cohosh or black snakeroot is a perennial herbaceous plant of the Ranunculaceae (buttercup family) with clusters of small white flowers. Collection from the wild mainly occurs in Illinois, Indiana, Kentucky, Missouri, Ohio, Tennessee, and West Virginia. Black cohosh has been used to treat a wide range of ailments, and is considered particularly effective to treat symptoms of menopause. Worldwide, sales of this species have exceeded those of goldenseal. Black cohosh is popular as a remedy in Europe, and it experienced a dramatic increase in consumption in the U.S. at the end of the twentieth century. It was identified as one of the fastest growing herbal products in 1998, with a more than 500 percent increase compared with previous year sales. Demand in the country is expected to continue increasing. There are also significant markets in China and Korea (Lyke 2001).

Plants for Ornamental and Landscape Use

The therapeutic values of plants extend beyond their direct medicinal values. It is increasingly recognized that plant-rich environments are important for human well-being, whether in the wild or in garden settings. Plant-rich environments are also important for the economy. The estimated annual value of aesthetic and passive use of forest ecosystem services is $280 million in the U.S. USFS (2019a).

Ornamental horticulture is important for well-being and also to the economy. The rich flora of the U.S. has provided many plants grown in gardens worldwide. Seventeenth century plant explorers sent species such as cutleaf coneflower (*Rudbeckia laciniata*), Virginia creeper (*Parthenocissus quinquefolia*), and tulip tree (*Liriodendron tulipifera*) back to Europe as exotic garden introductions. Botanists were active in gathering North American plants from the time of European colonization. The Gardeners' Company of London was among the companies that contributed financially to the founding of Virginia, motivated by the promise of novelties for the expansion of magnificent European gardens.

Native plants continue to be developed for commercial horticulture and there is growing interest in the use of native species for wildlife gardening and in landscaping. With over 79 million acres (32 million ha) of aesthetically managed land in the U.S., including urban and suburban landscapes and highway corridors, native plants provide major business opportunities. Native plants make up approximately 13 percent of the total sales of the nursery industry in the U.S. (Wilde et al. 2015).

North American plant species have been selected or bred for ornamental characteristics such as flowering, architecture, foliage, and disease-resistance traits. Examples include selection of native plant genotypes for early flowering in shrubby cinquefoil (*Potentilla fruticosa*), double flowers in redbud (*Cercis canadensis*), and compact form in oakleaf hydrangea (*Hydrangea quercifolia*). Most breeding and selection of North American plant species for ornamental traits has been conducted by private sector programs. Botanic gardens are also actively involved in this work. The Chicago Botanic Garden has, for example, been using wild collected germplasm and cultivated plants from nursery sources to produce improved garden plants of false indigo (*Baptisia*) and coneflowers (*Echinacea*) that are more suitable for the Midwest (Ault 2003).

Challenges remain in providing ornamental North American plants that are both ecologically functional and economically viable. Changes in flower or foliage color, petal doubling, and other traits may affect pollinator attraction, palatability, or resources available to wildlife who depend on native plants. Similar to environmental restoration with native plants as discussed in Chap. 6, attention needs to be paid to genetic diversity and local adaptability. As discussed by Wilde et al. (2015), a niche market has developed in the U.S. for locally adapted native ornamental plants that support wildlife and enhance biodiversity in urban areas. The single largest venture producing plants for this market is the American Beauties program, a partnership between the National Wildlife Federation and wholesale nurseries. Established in 2006, this program distributes over 350 native landscaping plant taxa to independent garden centers and landscapers in the northeastern U.S. Native plants for ornamental horticulture generally come from plant breeding, genotype selection, or open-pollinated seed. Within the American Beauties program, approximately 35 percent of the plants are named cultivars and 65 percent are propagated genotypes of native species.

Plants for the Economy

The contribution of wild plants to the agricultural economy is immense. The economic value of medicinal and ornamental plants in the U.S. is also significant. The contribution of native trees to the forestry economy is of major importance. The forest products industry accounts for approximately 4.5 percent of the total U.S. manufacturing gross domestic product, manufactures approximately $190 billion in products annually, and employs nearly 900,000 people (USFS 2019b). In 2015, the industrial roundwood production in the U.S. amounted to 14.1 billion cubic feet with industrial roundwood consumption in same year amounting to some 15.38 billion cubic feet (Statista 2019). The pulp and paper sector of the U.S. is the largest in the world and represents a considerable source of raw material for both the domestic and international markets. Most of the domestically produced fiber is sourced from naturally regenerated forests or planted native species.

Wood is not only important for construction, furniture, and fuel but for a surprising array of new industrial products. Cellulose, an important structural component of the cell walls of plants, is the most abundant natural polymer available. Nanocrystalline cellulose (NCC) is a new product, produced by processing wood pulp, including wood waste, a by-product of forestry. NCC is cheap to produce, light, strong, and conducts electricity. Some of its uses include computer components, lightweight body armor, and ballistic glass. The USFS opened the first pilot production plant producing NCC in the U.S. in Madison, Wisconsin in 2012. The U.S. National Science Foundation predicts NCC production will become a $600 billion industry by 2020 (Ferguson 2012).

Plants have the unique ability to convert solar energy to stored chemical energy by transforming atmospheric carbon dioxide and water into sugar molecules through the process of photosynthesis. Research is underway to artificially replicate this energy conversion process for the production of environmentally friendly and sustainable fuels, such as hydrogen and methanol. In the meantime, the search for alternatives to fossil fuels includes the use of a wide range of plant materials as a primary source of biomass. Bioenergy derived from biomass can be used as solid, liquid, and gaseous fuels for a wide range of uses, including transport, heating, electricity production, and cooking. The use of plant-based material to produce energy comprises about 28 percent of the U.S. renewable energy supply. Bioenergy from all sources in the U.S. could, in theory, supply nearly one-third of current national gasoline consumption, but this would depend on appropriate policies (Joyce et al. 2014). Bioenergy systems can cause both positive and negative environmental and social effects, depending on location of production and the technology, governance, and systems used. Low-carbon biofuels have clear potential to reduce greenhouse gas emissions, but the need to devote substantial land to their production raises challenging questions about how biofuel production will affect biodiversity (Evans et al. 2015). In the U.S., federal programs and private sector certification schemes require biofuel producers to consider such impacts. The use of forest residues from existing forestry operations is generally considered to be environmentally beneficial. The USFS is facilitating the use of renewable forest-based biomass, wood residues, and wood energy crops to contribute to national goals of 16 billion gallons (60.6 billion liters) of advance biofuels by 2022. Wood pellets made from highly compressed waste sawdust are becoming increasingly popular as a domestic fuel in the U.S. and are a significant export to Europe and other overseas markets.

Dedicated biomass plantations include native trees such as pine, willow (*Salix spp.*), and poplar (*Populus spp.*) together with annual crops and perennial plants such as switchgrass (*Panicum virgatum*). Perennial bioenergy plants are considered one of the most sustainable sources of renewable transportation fuel because they produce large amounts of biomass, require limited input of water and nutrients, and minimize ecological damage to soils and rivers. Switchgrass is a species of tallgrass prairies in the eastern U.S. Across its wide range, switchgrass populations exist either as upland or lowland ecotypes that differ in habitat preference, morphology, and productivity (Jager et al. 2010). Within the range of potential energy crops, more than 20 native species are considered important (Quinn et al. 2014). For example, plants of the succulent genus *Agave* have promise for production in hot and arid regions (Lewis et al. 2015).

In the U.S., future biomass production is likely to occur on land that is less suited to high-value row crop agriculture. It has been shown that the integration of bioenergy systems into agricultural landscapes can improve land and water use efficiency and help address concerns about environmental impacts (Creutzig et al. 2015). It has also been suggested that harvesting biomass directly from conserved prairie grasslands can be more beneficial than growing bioenergy crops as monocultures (Jungers et al. 2015).

Using water residues from natural forest management and carefully harvesting biomass from managed prairies are just two ways to generate financial income from plant diversity as well as conserving the land. In general, measuring the economic value of ecosystem services provided by plants can be difficult to quantify, but the value is estimated in trillions of dollars. The costs of ecological restoration to restore essential ecosystem services are discussed in Chap. 7. Despite the underlying importance of plants to the economy and well-being of the U.S. population, there is still a lack of complete knowledge of the flora and the plant species on which we depend. At the same time, the genetic diversity of native plant species is being eroded through the loss of wild populations locally, and an increasing number of plant species face extinction.

Plants Under Threat

Wild plants have a huge range of values, but there are usually major conflicting demands on the land where they grow, and many species are declining in the wild. Major continuing threats to the native flora of the U.S. include over-exploitation, habitat loss and fragmentation, and the impact of introduced invasive species. All of these factors have had a significant impact over the past four centuries. Pollution has grown in significance as a threat with increasing industrialization and use of agrochemicals and, more recently, global climate change has become an overarching threat. The full impact of climate change is only now being realized. As all plants, including widespread and naturally rare species, are affected by climate change, approaches to plant conservation need to evolve to predict the future as well as protecting remnants of the past.

Habitat Loss

Logging forests, clearing woodlands, plowing prairies, draining wetlands and overgrazing drylands have significantly changed the ecosystems of the U.S., as discussed in Chaps. 3 and 5. Habitat loss, whether for agricultural production, timber processing, mining, urbanization, transport, or other factors, has left only fragments of natural vegetation. Individual plant species cope with habitat loss in different ways depending on their overall distribution, ecology, and dispersal strategies. Altering or removing the underlying soil through mining, drainage, and compaction has a profound impact on native plant communities and individual species, not least by damaging the native seed bank. Naturally rare species confined to small areas are particularly vulnerable to habitat loss and may be particularly vulnerable to climate change. It has been noted by McKinney (2002) that urban development produces some of the greatest local extinction rates, and the impact is often more lasting than other types of habitat loss. In New England, for example, ecological succession is restoring forest habitat lost through farming and logging, but the spread of urbanization continues to threaten other habitat types in the region.

Invasive Species

A major threat to native plants, associated with modern agriculture (and other human activities), is the spread of invasive and noxious weeds. Over the past several centuries, more than 3,700 exotic plant species have been introduced, either accidentally or deliberately, into the U.S., with the largest concentrations occurring in and around large coastal cities (Kartesz 1999). Particularly rapid colonization by weedy introduced species occurs in habitats that have already been degraded by other factors such as logging, grazing, and burning. Globalization of goods and services and changing climate are increasing the range and diversity of weedy invasives and introduced pests and diseases that are impacting the native flora. It is estimated that invasive species are a threatening factor for 42 percent of U.S. federally listed Threatened and Endangered plant species and the main threat for 18 percent of these (Pimentel et al. 2005). The annual cost of invasive species to the U.S. economy has been estimated at $120 billion (Pimentel et al. 2005). One single species, the yellow starthistle (*Centaurea solstitialis*), native to the Mediterranean Basin and introduced to California in around 1850, currently costs ranchers and farmers $17 million in forage and control activities and $75 million in water losses annually (Groffman et al. 2014).

Box 2.3: Threats to a Keystone Conifer Species – Whitebark Pine

Whitebark pine (*Pinus albicaulis*) occurs in the western U.S. and Canada, with 90 percent of the species occurring on public lands in the two countries. It is an important keystone species in upper and subalpine ecosystems and helps to protect watersheds. Formerly, a very important economic species supporting the Montana mining industry, whitebark pine now faces four major and interlinked threats: the exotic fungus, white pine blister rust (*Cronartium ribicola*), the native mountain pine beetle (*Denroctonus ponderosae*), altered fire regimes, and climate change. The impacts of warming temperatures and decreased precipitation are predicted to cause a decline in suitable habitat; increased mountain pine beetle activity; increase in number, intensity, and extent of wildfires; and an increase in blister rust. Modeling projects indicate that whitebark pine could diminish to an area equivalent to less than 3 percent of its current distribution.

Fortunately, whitebark pine demonstrates ecological plasticity, and this can help buffer environmental changes. Phenotypic variation in whitebark pine is matched by moderate to high levels of genetic variation as shown by genetic studies. Gene conservation efforts are being undertaken by the USFS Forest Service in collaboration with the Bureau of Land Management (BLM), National Park Service (NPS), U.S. Fish and Wildlife Service (FWS), and agencies in Canada. These include the development and management of seed, pollen, and clone banks, together with seed orchards. Over 3,700 acres (1500 ha) of rust resistant seedlings have been planted and research is underway to augment natural regeneration of this important tree species that is listed as globally Endangered by International Union for the Conservation of Nature (IUCN).

Source: Mahalovich (2014)

Overexploitation

Overexploitation of plants of economic importance is a major threat to species around the world and surprisingly continues to be a problem for the U.S. flora because many species are important for rural livelihoods. Collection of rare species of cacti, orchids, and insectivorous plants continues to supply the demand for specialist horticulture and hobbyists. On a larger scale, commercial collection of medicinal plants from the wild, including species described in this chapter, both for domestic consumption and export overseas continues to require monitoring and regulation to prevent the loss of local populations and ultimately entire species. Harvesting of timber has also threatened the survival of certain tree species, at least at a local level.

Climate Change

Climate change is having a profound effect on plant diversity and significantly modifying characteristics of U.S. vegetation and habitats. Temperature and rainfall are fundamental to plant growth, and climatic factors are a major determinant of plant distribution. With climate change, plant ranges are shifting elevationally and latitudinally. Faced with climate change, species can adapt, migrate, or become extinct. Widespread species with greater genetic variation are able to fare better than naturally rare species that may also have a restricted genepool. Plants with restricted ranges, such as alpines, which are often isolated at higher elevations, may simply have nowhere to go. Despite current limitations in species distribution data and uncertainties in anticipated changes in rainfall and temperature, information on the predicted impact of climate change on rare plants and more widespread species is essential for managing plants.

U.S. forests are expected to experience extensive ecological, social, and economic impacts as a result of climate change (Malmsheimer et al. 2008), with their ecosystem functions and attributes significantly altered. Climate change will pose a severe threat to the viability of forest tree species, if they are unable to adapt to new conditions or to shift their ranges to more favorable environments. Evidence suggests that tree species are already exhibiting changes in distribution and phenology in response to climate change. Within the Appalachian Mountains, three forest tree taxa likely to be at greatest risk of extinction as a result of climate change are Fraser fir (*Abies fraseri*), listed as Endangered by IUCN; intermediate balsam fir (*Abies balsamea* var. *phanerolepis*), listed as Data Deficient by IUCN; and Carolina hemlock (*Tsuga caroliniana*), listed as Near Threatened. These conifers are limited to small, high-elevation populations, and are under threat from exotic pests: balsam woolly adelgid (*Adelges piceae*), which impacts the two fir (*Abies*) taxa, and hemlock woolly adelgid (*Adelges tsugae*), which attacks Carolina hemlock (Potter et al. 2010).

Global climate change is already resulting in increased frequency of extreme weather events. Drought has increased significantly in the West, with impacts on desert, woodland, and forest vegetation. Another factor associated with global climate change is increased fire frequency and intensity. Between 1980 and 2011, the average annual number of fires more than doubled on all federal lands, and the total area burned in a typical year has more than tripled to over 3 million acres (1.2 million ha). Since 2000, there have been 4 years where more than 6 million acres (2.4 million ha) burned on federal lands in the West, whereas the 20 years prior to that saw only 2 years where total burned area exceeded 4 million acres (1.6 million ha). The long-term trend demonstrates that the average year now has twice as many large fires as 40 years ago, and the likelihood of a particularly devastating fire year is growing rapidly. Arizona, California, Colorado, Idaho and Montana have seen the most dramatic increases in wildfires since 1970 (Climate Central 2012).

Wildfire risk depends on factors such as temperature, soil moisture, and the condition of vegetation and other potential sources of fuel. All these factors are linked directly or indirectly to climate change.

Where are the Plant Diversity Hotspots?

The intensity of threats to wild plant species varies geographically as does the level of diversity of native species. The rich and diverse flora of the U.S. is uniquely adapted to local environmental conditions. Some parts of the nation have an extraordinary range of species with many local endemics. These plant diversity hotspots have attracted particular research and conservation attention.

On a world scale California, Hawaii, and the North American Coastal Plain are recognized as globally important biodiversity hotspots. The California Floristic Province is a region of extraordinary plant richness. This region extends over approximately 120,000 square miles (310,000 square km) and is defined by its Mediterranean climate with hot, dry summers and cool, wet winters, and by common ecological features. With elevations ranging from 660 feet (200 m) below sea level in Death Valley to about 13,000 feet (4,000 m), the Province includes almost all of the State of California, except its deserts and the northeastern Modoc Plateau, as well as adjacent parts of Mexico and Oregon.

The California hotspot owes its plant diversity to the wide range of ecological conditions that are reflected in its diverse ecosystems, which include sagebrush steppe, chaparral, pinyon-juniper woodland, and a wide range of forest types, coastal dunes, and salt marshes. Today, about 30,000 square miles (80,000 square km) or 24.7 percent of the original vegetation, remains in more or less unaltered condition. There are over 2,600 species of vascular plants in the hotspot, with 42 percent found nowhere else in the world (Burge et al. 2016). The high levels of plant endemism are due to the varied topography, climate zones, geology, and soils.

Within this extraordinary hotspot, areas of exceptionally high plant diversity are the Sierra Nevada Mountains, the Transverse Ranges in southern California, the Klamath-Siskiyou region in the coastal mountain ranges of California and Oregon, and the Coast Ranges. The Transverse Ranges, forming a narrow strip that runs east to west in southern California, separate the Coast Ranges to the north from the Peninsular Ranges to the south. The Klamath-Siskiyou region bridges the coastal mountain ranges of California and Oregon, and is home to the most diverse temperate coniferous tree community in the world.

In addition, serpentine soil habitats occur along fault zones in the Central and North Coast and Cascade ranges, from sea level to an elevation of 9,500 feet (2,900 m). Serpentine habitats have nutrient-poor soils which support a highly specialized and diverse flora. It has been estimated that serpentine endemic plant species represent 10 percent of the California Floristic Province's endemics.

The hotspot is also home to two spectacular endemic tree species, the giant sequoia (*Sequoiadendron giganteum*) and the coastal redwood (*Sequoia sempervirens*). The giant sequoia, which remains in 75 groves in the Sierra Nevada range, is the most massive species ever to live on Earth, reaching heights of 250 feet (75 m) and circumferences of 98 feet (30 m) in the oldest trees. The closely related redwood is often even taller, sometimes reaching 345 feet (105 m), although smaller in circumference.

Human population pressures have rendered California one of the four most ecologically degraded states in the U.S. (see Box 5.3. in Chap. 5). Among the states with the greatest number of plants at risk, California is in second place with Hawaii having the most threatened plants (Stein 2002). Nevertheless, conservation efforts are impressive as described in Chap. 4. Approximately 37 percent of the total land area of the California Floristic Province is under official protection.

Hawaii has an extraordinary flora and one that is under great threat. Located in the Pacific Ocean over 2,000 miles (3,200 km) from the nearest continental land mass, the islands of Hawaii are the most isolated high islands in the world. This extreme isolation, together with a high diversity of habitat types, helps to explain the uniqueness of the flora, with more than 90 percent of the native plants being endemic. In total, about 1,500 plant species are native to Hawaii and around 20 of these have been newly described in the past two decades. Many of these are threatened with extinction as a result of deforestation, the effects of fires, overgrazing by introduced livestock, and competition with invasive plant species. Hawaii has nearly half of the plant extinctions in the U.S. over the past 30 years.

The North American Coastal Plain hotspot includes a small section of northern Mexico and extends along the Gulf of Mexico, through Florida and up the East Coast to southeastern Massachusetts (Noss et al. 2015). This region was classified as a global biodiversity hotspot in 2016. Texas, a state rich in biological diversity, ranking highly in diversity, endemism, and number of extinctions, falls partly within the North American Coastal Plain hotspot. This huge state overlaps several major ecological regions, including the southwestern deserts, the Great Plains, the humid Gulf Coast, and, at the state's southern tip, the Mexican subtropics. As a result, many species reach distributional limits in Texas, and a unique blend of eastern and western species occurs within the state. Distinctive landforms, including the limestone Edwards Plateau in Central Texas, support rich assemblages of biodiversity with some of the rarest species in the U.S.

The Edwards Plateau (Ecoregion 7) has an area of nearly 24 million acres (9.7 million ha). This semi-arid rugged landscape which experiences long periods of drought has vegetation dominated by Ashe's juniper (*Juniperus ashei*), oaks, and honey mesquite (*Prosopis glandulosa*). Much of the region overlays the immense underground reservoir of the Edwards Aquifer that feeds crystal clear streams. More than 100 of the 400 endemic plants of Texas occur within the Edwards Plateau. These include the shrubs, Texas snowbells (*Styrax texana*), a beautiful plant that is endangered, and Texabama croton (*Croton alabamensis var. texensis*) together with bracted twist-flower (*Streptanthus bracteatus*), and Texas wildrice (see Box 2.2). In protected valleys there are disjunct populations of Texas madrone (*Arbutus xalapensis*), Texas smoke tree (*Croton obovatus*), witch hazel (*Hamamelis* spp.) and big-tooth maples (*Acer grandidentatum*). The moist river corridors are lined with bald cypress (*Taxodium* spp.), pecan (*Carya* spp.), hackberry (*Celtis* spp.), and sycamores (*Platanus* spp.). This region is also host to some of the most spectacular wildflower displays featuring bluebonnets, Indian paintbrush, gaillardia (*Gaillardia* spp.), and golden-wave (*Coreopsis* spp.).

The conservation importance of national and global botanical hotspots cannot be underestimated. However, valuing and appreciating local native plant diversity is important throughout the country both in rural and urban situations. It is the local native plant diversity that gives us our sense of place.

Box 2.4: Saguaro – An Iconic Keystone Species

The iconic saguaro (*Carnegiea gigantea*) is only found in the Sonoran Desert, extending from Arizona south through Sonora, Mexico. It is a keystone species of the Sonoran Desert providing ecosystem services for other plants and animals and a wide variety of products for human use. The saguaro is pollinated by bats and probably moths during the night and by birds, including the white-winged dove (*Zenaida* asiatica), bees, and other insects by day. The white-wing dove and elf owl (*Micrathene whitneyi*) share the same area of distribution with saguaro on which they depend. The dove and many other birds feed on the fruit, whereas the tiny elf owl nests in abandoned woodpecker holes in the cactus trunks. Native Americans use the fruit for making drinks and jellies, and the wood is used for construction and handicrafts. Medicinal products are also derived from the plant. The majestic saguaro, so characteristic of the desert landscape, is threatened by grazing, competition with invasive species such as buffelgrass (*Pennisetum ciliare*), poaching, and urban expansion. For many decades this species has declined, however, census information from 2010 indicates that recruitment, as measured by the number of small saguaros, was substantially higher than at previous times.

Source: O'Brien et al. (2011)

References

Ault J (2003) Breeding and development of new ornamental plants from North American native taxa. In: XXVI International Horticultural Congress: Elegant Science in Floriculture. Acta Horticulturae 624:37–42

Burge DO, Thorne JH, Harrison SP, O'Brien BC, Rebman JP, Shevock JR, Alverson ER, Hardison LK, Rodriguez JD, Junak SA, Oberbauer TA, Riemann H, Vanderplank SE, Barry T (2016) Plant diversity and endemism in the California Floristic Province. Madrono 63(2):3–206. https://doi.org/10.3120/madr-63-02-3-206.1

Calderone NW (2012) Insect pollinated crops, insect pollinators and US agriculture: trend analysis of aggregate data for the period 1992–2009. PLoS One 7(5):e37235. https://doi.org/10.1371/journal.pone.0037235

CBD (2019) Texas wild rice. https://www.biologicaldiversity.org/campaigns/esa_works/profile_pages/TexasWildRice.html. Accessed 11 June 2019

CBWF (2019) Crested Butte Wildflower Festival. https://www.crestedbuttewildflowerfestival.com. Accessed 11 June 2019

Chamberlain JL, Prisley S, McGuffin M (2013) Understanding the relationships between American ginseng harvest and hardwood forests inventory and timber harvest to improve co-management of the forests of Eastern United States. J Sustain For. https://doi.org/10.1080/10549811.2013.798828

Clark J (2017) Center for Plant Conservation President's Message. https://saveplants.org/about-us/president-message/. Accessed 25 Feb 2019

Climate Central (2012) The age of Western wildfires. http://www.climatecentral.org/wgts/wildfires/Wildfires2012.pdf. Accessed 3 Apr 2018

Creutzig F, Ravindranath NH, Berndes G, Bolwig S, Bright S, Cherubini F, Chum H, Corbera E, Delucchi M, Faaij A, Fargione J, Haberl H, Heath G, Lucon O, Plevin R, Popp A, Robledo-abad C, Rose S, Smith P, Stromman A, Sanngwon S, Masera O (2015) Bioenergy and climate change mitigation: an assessment. GCB Bioenergy 7(5):916–944. https://doi.org/10.1111/gcbb.12205

Evans SG, Kelley LC, Potts MD (2015) The potential impact of second-generation biofuel landscapes on at-risk species in the US. GCB Bioenergy 7(2):337–348. https://doi.org/10.1111/gcbb.12131

Ferguson W (2012) Why wood pulp is world's new wonder material. https://www.newscientist.com/article/mg21528786–100-why-wood-pulp-is-worlds-new-wonder-material/ Accessed 3 May 2018

FWS (2013) U.S. exports of American ginseng 1992–2012. https://www.fws.gov/International/pdf/report-us-exports-of-american-ginseng-1992-2012.pdf. Accessed 5 June 2019

FWS (2015) Wild American ginseng: information for dealers and exporters Item FW 5001. https://fws.gov/international/pdf/factsheet-american-ginseng-harvesters-dealers-exporters.pdf. Accessed 5 June 2019

Gibbons B (2011) Wildflower wonders: The 50 best wildflower sites in the world. Bloomsbury, London, UK, 192 p

Greene S, Khoury C, Williams K, Wiersema J, Kinard G (2015) Collecting crop wild relatives: an emerging priority. Paper presented at the 2015 National Native Seed Conference, Santa Fe, New Mexico, 13–16 April 2015

Groffman PM, Kareiva P, Carter S, Grimm NB, Lawler J, Mack M, Matzek V, Tallis H (2014) Ecosystems, biodiversity and ecosystem services. In: Melillo JM, Richmond TC, Yohe GW (eds) Climate change impacts in the United States: the third national climate assessment. US Global Change Research Program, pp 195–219. https://doi.org/10.7930/J0TD9V7H http://nca2014.globalchange.gov/report/sectors/ecosystems Assessed 3 Apr 2018

Hatfield J, Takle G, Grotjahn R, Holden P, Izaurralde RC, Mader T, Marshall E, Liverman D (2014) Ch. 6: Agriculture. In: Melillo JM, Richmond TC, Yohe GW (eds) Climate change impacts in the United States: the third national climate assessment. U.S. Global Change Research Program, pp 150–174. https://doi.org/10.7930/J02Z13FR

Jager HI, Baskaran LB, Brandt CC, Davis EB, Gunderson CA, Wullschleger SD (2010) Empirical geographic modeling of switchgrass yields in the United States. GCB Bioenergy 2(5):248–257. https://doi.org/10.1111/j.1757-1707.2010.01059.x

Joyce LA, Running SW, Breshears DD, Dale VH, Malmsheimer RW, Sampson RN, Sohngen B, Woodall CW (2014) Forests. In: Melillo JM, Richmond TC, Yohe GW (eds) Climate change impacts in the United States: the third national climate assessment. https://doi.org/10.7930/J0Z60KZC

Jungers JM, Sheaffer CC, Fargione J, Lehman C (2015) Short-term harvesting of biomass from conservation grasslands maintains plant diversity. GCB Bioenergy 7:1050–1061

Kartesz J (1999) Exotic vascular plant species: Where do they occur? In: Ricketts TH, Dinertein E, Olson DM et al (eds) Terrestrial ecoregions of North America a conservation assessment. Island Press, Washington, D.C, p 76–81

Khoury CK, Greene S, Wiersema J, Maxted N, Jarvis A, Struik P (2013) An inventory of crop wild relatives of the United States. Crop Sci 53(4):1496

Lewis SM, Gross S, Visel A, Kelly M, Morrow W (2015) Fuzzy GIS-based multi-criteria evaluation for US Agave production as a bioenergy feedstock. GCB Bioenergy 7:84–99. https://doi.org/10.1111/gcbb.12116

Lyke J (2001) Conservation status of *Cimicifuga rubifolia*, *C. americana*, and *C. racemosa*. Med Plant Conserv 7:22–24

Mahalovich M (2014) Effects of pests and diseases on *Pinus albcaulis* in the USA. Case study. In: Shaw K, Hird A (eds) Global survey of ex situ conifer collections. BGCI, Richmond, UK

Malmsheimer RW, Heffernan P, Brink S, Crandall D, Deneke F, Galik C, Gee E, Helms JA, McClure N, Mortimer M, Ruddell S, Smith M, Stewart J (2008) Forest management solutions for mitigating climate change in the United States. J For 106(3):115–173

McKinney ML (2002) Urbanization, biodiversity, and conservation. Bioscience 52(10):883–890

Noss RF, Platt WJ, Sorrie BA, Weakley AS, Means DB, Costanza J, Peet RK (2015) How global biodiversity hotspots may go unrecognized: lessons from the North American Coastal Plain. Divers Distrib 21:236–244

O'Brien K, Swann DE, Springer AC (2011) Results of the 2010 Saguaro Census at Saguaro National Park. https://ww.nps.gov/sagu/learn/nature/upload/Saguaro-Census-2011.pdf. Accessed 25 Feb 2019

Pimentel D, Zuniga R, Morrison D (2005) Update on the environmental and economic costs associated with alien-invasive species in the United States. Ecol Econ 53(3):273–288

Potter KM, Hargrove WW, Koch FH (2010) Predicting climate change extirpation risk for central and southern Appalachian forest tree species. In: Rentch JS, Schuler TM (eds) Proceedings from the conference on the ecology and management of high-elevation forests in the Central and Southern Appalachian Mountains. Snowshoe, West Virginia, May 14–15, 2009. General Technical Report NRS-P-64. U.S. Department of Agriculture, Forest Service, Northern Research Station, Newtown Square, PA, pp 179–189

Quinn LD, Gordon DR, Glaser A, Lieurance D, Flory SL (2014) Bioenergy feedstocks at low risk of invasion in the USA: a "white-list" approach. Bioenergy Res. https://doi.org/10.1007/s12155-014-9503-z

Richards CM, Antolin MF, Reilly A, Poole J, Walters C (2007) Capturing genetic diversity of wild populations for ex situ conservation: Texas wild rice (*Zizania texana*) as a model. Genet Resour Crop Evol 54(3):837–848

Sari AO, Morales MR, Simon JE (1999) *Echinacea angustifolia*: an emerging medicinal. In: Janick J (ed) Perspectives on new crops and new uses. ASHS Press, Alexandria, VA, pp 490–493

Shevock J, Taylor DW (1988) Plant exploration in California, the frontier is still here. In: Elias T (ed) Conservation and management of rare and endangered plants. Proceedings of California conference on the conservation and management of rare and endangered plants. California Native Plant Society, pp 91–98

Statista (2019) U.S. forest products industry - statistics & facts. https://www.statista.com/topics/1316/forest-products-industry/. Accessed 11 June 2019

Stein B (2002) States of the Union: ranking America's biodiversity. NatureServe, Arlington, VA, 25 p

USDA (2014) Joint strategic framework on the conservation and use of native crop wild relatives in the United States. FS-1029. USDA Forest Service and Agricultural Research Service, Washington, DC, 16 p

USFS (2019a) Ecosystem services from National Grasslands. https://www.fs.fed.us/grasslands/ecoservices/index.shtml

USFS (2019b) Energy & Forest Products. https://www.fs.fed.us/science-technology/energy-forest-products. Accessed 11 June 2019

Wilde HD, Gandhi KJ, Colson G (2015) State of the science and challenges of breeding landscape plants with ecological function. Hortic Res 2:1–8. https://doi.org/10.1038/hortres.2014.69

World Resources Institute (WRI) (2008) Ecosystem services: a guide for decision makers. World Resources Institute, Washington, D.C.

Chapter 3
The Importance of Ecosystems

I love all of the ecosystems – mountains, deserts, rainforests. They're beautiful, and nature has so many different flavors to it. (Louie Schwartzberg 2014)

The rich botanical diversity of the U.S. is reflected in the diverse landscapes, vegetation types, and ecosystems of the nation. In natural areas, plants grow in different assemblages or communities, interacting with other species and physical factors to form ecosystems. Current ecosystems are a product of climatic and edaphic factors, together with historical management and natural disturbance regimes. Natural ecosystems form the country's green infrastructure, providing essential ecosystem services such as carbon storage, regulation of water supplies, flood control, and erosion control. Healthy ecosystems provide ecological stability and resilience by supporting livelihoods and buffering against adverse impacts of environmental change.

In the U.S., as elsewhere around the world, terrestrial ecosystems are inherently variable, but for management, mapping, and monitoring purposes they are generally identified and distinguished on the basis of plant communities - areas with similar plant species composition and structure. The dominant trees, grasses, forbs, and their assemblages reflect the underlying geology, soils, and climatic conditions. The U.S. National Vegetation Classification, Version 2, adopted by the Federal Geographic Data Committee (2008), provides a standard approach to the description of plant species assemblages. This allows consistency in naming and mapping vegetation across the country and by different agencies.

For mapping purposes, ecosystems are frequently aggregated into ecoregions. The North American ecoregion classification system commonly used in the U.S. has 12 broad, Level I ecological regions based on Omernik (1987). These incorporate spatial variation in climate, soil, geology, vegetation, and physiography and relate well to what can be seen on the ground. The Level I Ecoregions (see Fig. 3.1) delimit major ecological areas and provide the broad backdrop to the ecological mosaic of the continent, putting it in context at global or intercontinental scales. For convenience we think in broad terms such as forests, prairies, and deserts, which equate

S. F. Oldfield et al., *Seeds of Restoration Success*, Springer Earth System Sciences, https://doi.org/10.1007/978-3-319-96974-9_3

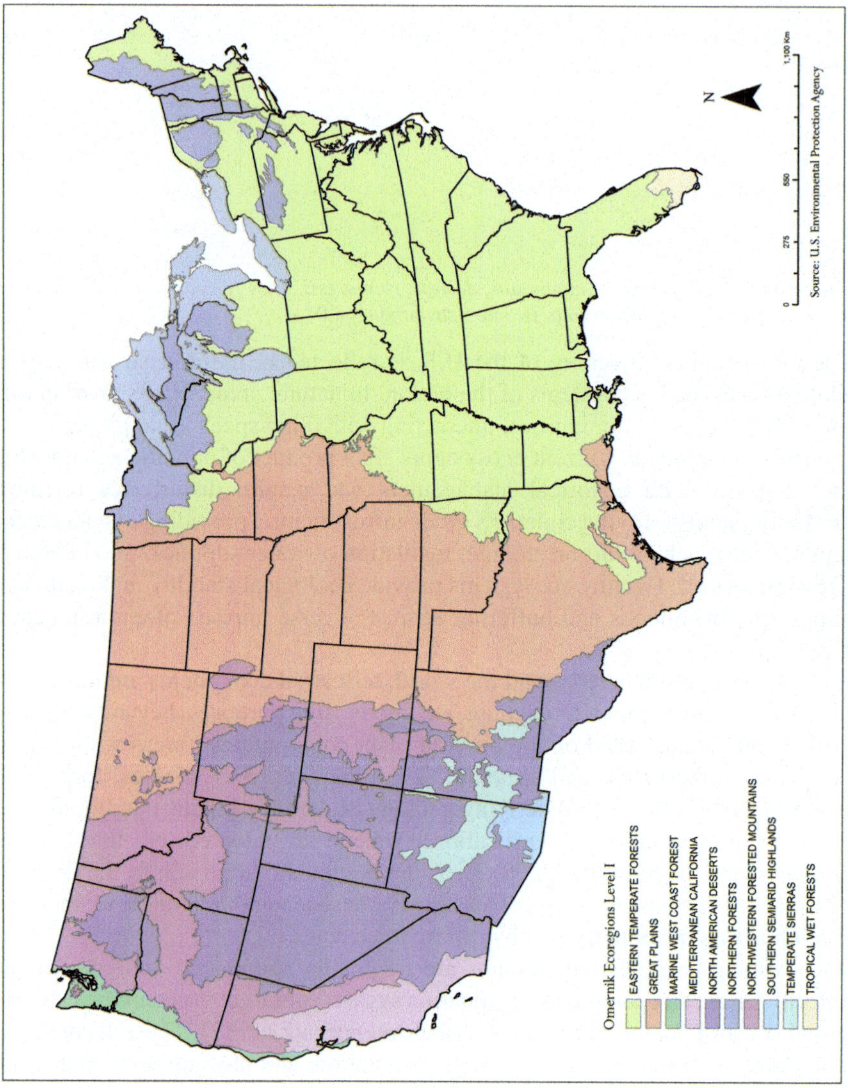

Fig. 3.1 Omernik Level I Ecoregions in the U.S. (Omernik 1987)

broadly to Level I Ecoregions. In some cases the Ecoregions correspond more clearly to a relatively uniform vegetation type as, for example, the Eastern Temperate Forests Ecoregion or the Great Plains Ecoregion which correspond to the broad region of the prairies. In the West, where there is much more ecological complexity, each Level 1 Ecoregion covers a broad range of ecosystem types.

Finer level Ecoregions are useful for planning landscape level conservation and restoration activities for assessment, monitoring, and reporting. The 24 Level II Ecoregions are being used, for example, as the assessment and reporting units for the national assessment of carbon stocks and sequestration that was mandated by the 2007 Energy Independence and Security Act. Level III Ecoregions are used for conservation and ecological restoration planning and are the basis for Rapid Ecoregional Assessments undertaken by the Bureau of Land Management (BLM) in the West. Level III Ecoregions will be referred to in later chapters of this book, particularly in Chap. 6 where developing native seed for ecological restoration is discussed and a map of these Ecoregions is provided.

The following account provides a generalized description of major ecosystem types broadly relating to Level 1 Ecoregions as summarized by the Commission for Environmental Cooperation (1997) combined with information from the account of North American Terrestrial Vegetation provided by Barbour and Billings (2000). A map of major vegetation types based on Barbour and Billings (2000) is given in Fig. 3.2. Information on the current extent of major ecological units – forests, woodlands, prairies, deserts, and wetlands – comes from the latest available source material as indicated in the text. Information on their status draws on the assessment of terrestrial ecosystems of North America by Ricketts et al. (1999).

Forests

Forested land in the U.S. covers about 766 million acres (310 million ha) with about 55 percent in the East and 45 percent in the West (Oswalt et al. 2014). The predominantly temperate forests in mainland U.S. vary in composition at a landscape level according to current climate and vegetation history, reflecting climatic conditions in the ancient past and recent land use. Altitude and geology account for more local variation. Within the Level I Ecoregions forests of the U.S. are broadly divided into the Northern Forests, Eastern Temperate Forests, Tropical Humid Forest, Northwestern Forested Mountains, and Marine West Coast Forest.

The Northern Forests Ecoregion consists mainly of extensive forest areas in Canada with a relatively small portion in the U.S. The climate is characterized by long, cold winters and short, warm summers. The continental climate is influenced by maritime conditions near the coast and by cold arctic air masses from the north. The Great Lakes have a moderating effect on adjacent lands, providing warmer conditions in winter and cooling the local climate in summer. The portion of the Northern Forests Ecoregion falling within the U.S. consists of coniferous trees, largely white spruce (*Picea glauca*), black spruce (*Picea mariana*), jack pine (*Pinus banksiana*), balsam fir (*Abies balsamea*), and tamarack (*Larix laricina*).

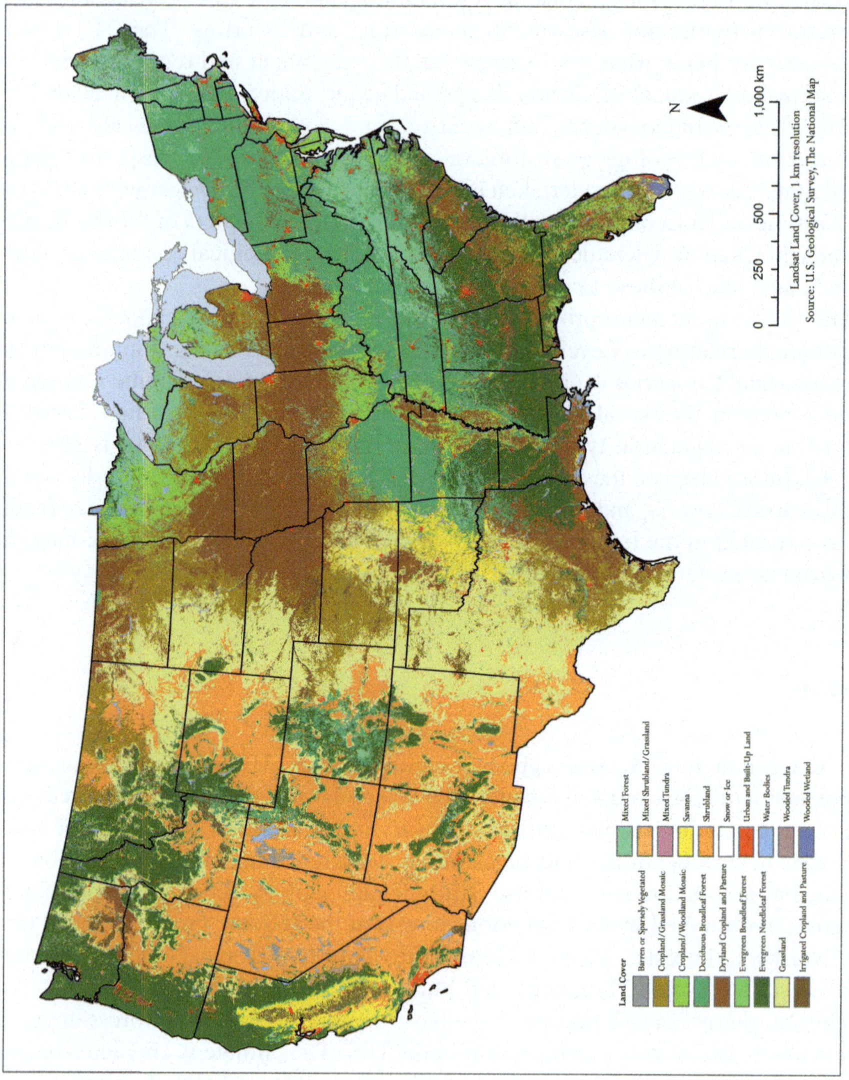

Fig. 3.2 Major vegetation types in the U.S. (Barbour and Billings (2002)

Other conifers include white pine (*Pinus strobus*), red pine (*Pinus resinosa*), and red spruce (*Picea rubens*). Common deciduous trees include white birch (*Betula papyrifera*), trembling aspen (*Populus tremuloides*), balsam poplar (*Populus balsamifera*), sugar maple (*Acer saccharum*), American beech (*Fagus grandifolia*), and oaks (*Quercus* spp.). The Northern Forest ecosystems found in the U.S. remain relatively unchanged compared to forests elsewhere in the country, but they have become fragmented due to logging, farming, and settlement.

The Eastern Temperate Forests Ecoregion is the largest forested ecoregion falling within the U.S. Temperate hardwood forests were the dominant land cover in the eastern U.S. prior to European settlement, forming a largely unbroken canopy from the Central Atlantic Coastal Plain to the eastern edge of the Great Plains (Delcourt and Delcourt 2000). The Eastern Temperate Forests still cover a huge area from New England to Florida and west to the prairies, interlocking with the Northern Forests Ecoregion to the north.

Eastern deciduous forests are characterized by a mild to warm, temperate and humid climate. The dry Pacific air mass, originating in the North Pacific Ocean, dominates in autumn and winter, corresponding with the dormant period of deciduous trees, whereas the tropical air mass brings wetter conditions in the spring and summer periods of growth. Forest cover in this Ecoregion has remained relatively constant over the past 100 years, but the composition of forests has changed with land use practices over time. The deciduous forest is broadly characterized by canopy dominance of three tree genera: oak, hickory, and basswood (*Tilia* spp.) together with three additional species tulip tree, American beech, and sugar maple. Red maple (*Acer rubrum*) has become increasingly abundant in the Eastern temperate forests, spreading from its original riparian habitats and causing some concern as it outcompetes oaks and hickories. Important conifers are eastern hemlock and white pine. The Eastern deciduous forests typically have a rich understory vegetation with a notable assemblage of spring-flowering herbaceous plants. Beech-maple and maple-basswood forest types occur widely, particularly in the eastern parts of this region, mixed oak-hickory associations are common in the Upper Midwest, changing into oak-hickory-pine mixed forests in the south and the Appalachians. Oaks form about 80 percent of the composition in the forests from the Southern Appalachians to the Ozark Highlands and the tulip tree locally accounts for 20–30 percent in this same area.

American chestnut (*Castanea dentata*) was formerly dominant in the Appalachian forests but was devastated by the introduction from Europe of Cryphonectria parasitica, formerly *Endothia parasitica*, a fungal blight, in the early part of the twentieth century (see also Chap. 6). Forest composition is continuing to change due to more recently introduced pests and pathogens, such as the Emerald Ash Borer (*Agrilus planipennis*). This beetle, accidentally introduced from Asia, is having a devastating impact on widespread North American ash species, such as white ash (*Fraxinus americana*), which is native to eastern and central U.S. White ash was common and widespread, although it was never a dominant species in the forest. It is now listed as Critically Endangered by IUCN because of the impact of Emerald Ash Borer.

Floristically the Eastern deciduous forests are closely related to temperate forests of eastern China, Japan, and parts of Europe with a shared history dating back to the Tertiary and with related groups of relict plants. Over 50 genera including magnolia, maple, hickory, stewartia, diospyros, and dogwood (*Cornus* spp.) and some species have disjunct distributions in these distant regions.

Box 3.1: The Appalachian Forests
The Appalachian forests extending from southern New York to Georgia, are among the most diverse temperate deciduous forests in the world, with only the temperate forests of China considered slightly richer. More than 150 tree species grow in the region, and the understory vegetation is also highly diverse. Forest communities often support more than 30 canopy tree species at a single site together with rich understories of ferns, fungi, perennial and annual herbaceous plants, shrubs, small trees, and diverse animal communities. Songbirds, salamanders, land snails, and beetles are examples of some particularly diverse groups of animals.

Over 80 percent of the habitat of the Appalachian mixed forests has been altered, particularly in the ridge and valley province. The majority of the more or less intact forests are public lands.

Source: Ricketts et al. (1999)

Also forming part of the Eastern Temperate Forests Ecoregion, and contributing to the North American Coastal Plain biodiversity hotspot (see Chap. 2), longleaf pine (*Pinus palustris*) ecosystems once occupied 94 million acres (38 million ha) in the southeastern U.S. and were among the most extensive ecosystems of North America. This broad range of ecosystems encompasses forests, woodlands, and savannas on a variety of sites ranging from wet flatwoods to xeric sandhills, and rocky mountainous ridges. Longleaf pine ecosystems occur from the Atlantic Ocean to the Great Plains and south to the Gulf of Mexico. Characterized by an open, parklike structure, longleaf pine ecosystems are a product of frequent fires, facilitated by the presence of fallen pine needles and bunchgrasses in the understory (Brockway et al. 2006).

The portion of the Tropical Humid Forest Ecoregion falling within the U.S. covers a relatively small area of southern Florida. The climate of this ecoregion is generally warm, with high rainfall throughout the year. Pine forests cover thin limestone soils with patches of tropical forest known as hardwood hummocks growing on slightly elevated ground. Tropical tree species such as mahogany (*Swietenia mahogoni*), gumbo limbo (*Bursera simaruba*), and cocoplum (*Chrysobalanus icaco*) grow among species with distributions that extend further north such as live oak (*Quercus virginiana*), red maple, and hackberry. Collectively these forests form rockland ecosystems. Only 2 percent of this habitat remains, with for example, an area within the Everglades National Park (Ricketts et al. 1999). Characteristic mangrove vegetation also occurs within the Everglades.

The Northwestern Forested Mountains Ecoregion corresponds broadly to the northern Rockies. This ecoregion has a diverse mosaic of ecosystem types, with dense conifer forest interspersed with alpine tundra at high altitudes together with dry sagebrush communities and areas of grassland. The forests throughout the Rockies have been significantly influenced by fire and are generally considered to be disturbance forests. Tracts of forest were traditionally burned by Native Americans with increased burning by settlers in the latter part of the nineteenth century, followed by cessation of burning in the twentieth century (Peet 2000). Lodgepole pine (*Pinus contorta*), subalpine fir (*Abies lasiocarpa*), silver fir (*Abies amabilis*), grand fir (*Abies grandis*), and Engelmann spruce (*Picea engelmannii*) are among the important conifers. With decreasing elevation, the vegetation of the mountain slopes and rolling plains changes to forests characterized by ponderosa pine (*Pinus ponderosa*), the species most commonly associated with western mountains, and Rocky Mountain or Interior Douglas fir (*Pseudotsuga menziesii* var. *glauca*). Riparian forests of the Rocky Mountain foothills have similarities in tree genera and goundcover species composition with the Eastern deciduous forests.

The Marine West Coast Forest Ecoregion has magnificent temperate rain forest stretching in a relatively narrow band, within 37–75 miles (60–120 km) from the Pacific Ocean, extending from northern California and southern Oregon to the Sitka spruce forests of southeast Alaska. These forests experience the wettest climates of North America with mild, moist maritime conditions. The Marine West Coast Forest ecoregion is characterized by north-south mountain ranges – the Cascade Range, Coastal Ranges of Oregon and Olympic Mountains. The dominant trees include coast Douglas fir (*Pseudotsuga menziesii* var.*menziesii*), western hemlock (*Tsuga heterophylla*), western red cedar (*Thuja plicata*), Sitka spruce (*Picea sitchensis*), and coast redwood (*Sequoia sempervirens*). The forests have been highly important for timber production with forestry being the major resource activity together, increasingly, with recreation. Many of these trees reach a huge size, coast redwood being the tallest tree species globally and coast Douglas fir the second tallest. Both live to a great age, forming ancient or old-growth forest.

Box 3.2: Forest or Woodland?

It has been estimated that the original forests of the U.S. covered 850 million acres (344 million ha) with an additional 100 million acres (40.5 million ha) of scrubby forest and brush land, chiefly in the West. The distinction between forest and woodland has been subject to much debate. The species involved (predominantly junipers, mesquite, and scrub oaks) tend to have a continuum of life forms ranging from tree to shrub, matching the transition from forest land to shrubland with both being matters of subjectivity. In the Midwest, forest and woodland are distinguished by tree density. Forests are the most dense, woodlands more open, and savannas still more open with scattered trees growing among the grass and forb understory. Recently, the USFS adopted new standards for what qualifies as forest. Ecosystems such as chaparral and sparse pinyon-juniper "forests" with 5–10 percent tree cover in the Rocky Mountains and California have been reclassified as woodlands rather than forestlands. The USFS estimates that the new definition removes about 6 million to 8 million acres (2.4 to 3.2 million ha) previously considered forestlands. Generally, these changes do not affect historic estimates of productive, unreserved forestland or timberland.

Source: Oswalt et al. (2014)

Woodlands

Woodlands of the U.S. are estimated to cover around 52.6 million acres (21.3 million ha) of land (Oswalt et al. 2014). Types of woodland include the pinyon-juniper woodlands of the American Deserts and California Ecoregions, chaparral and woodlands of the Mediterranean California Ecoregion, and oak savannas of the Great Plains Ecoregion.

Pinyon-juniper woodland extends across a huge area of the Southwest, occurring in 10 states and covering as much as 15 percent of the land area of Arizona, Colorado, Nevada, New Mexico, and Utah. This type of woodland is also found on the Edwards Plateau of Central Texas where mesquite-acacia savanna also occurs. Five common species of juniper are associated with pinyon-juniper woodland: Rocky Mountain juniper (*Juniperus scopulorum*), redberry juniper (*J. coahuilensis*), alligator juniper (*J. deppeana*), one-seeded juniper (*J. monosperma*), and Utah juniper (*J. osteosperma*). The commonly occurring pinyon species include common or two-needle pinyon (*Pinus edulis*), singleleaf pinyon (*Pinus monophylla*), and Mexican pinyon (*Pinus cembroides*). Pinyon-juniper woodland generally is found in regions subject to temperature extremes and limited moisture availability. Usually occurring at altitudes between 4,500 to 7,500 feet (1,370 and 2,290 meters), pinyon-juniper woodland is transitional from grasslands or shrublands at lower elevations, to ponderosa pine or other montane forest associations at higher elevations (Gori and Bate 2007). Pinyon-juniper woodland is expanding as a consequence of climate change.

Chaparral or sclerophyllous shrubland is one of the most extensive vegetation types of the Mediterranean California Ecoregion. The climate of this Ecoregion is characterized by hot, dry summers and mild winters, with precipitation associated with winter frontal storms from the Pacific. Chaparral is fire-adapted vegetation with plants that characteristically have thickened, hardened leaves resistant to water loss, forming a cover of closely spaced shrubs that grow up to 13 feet (4 m tall). Common shrubs include chamise (*Adenostoma fasciculatum*), buckbrush (*Ceanothus cuneatus*), and manzanita (*Arctostaphylos* spp.). Conifers, such as knobcone pine (*Pinus attenuata*), big-cone Douglas fir (*Pseudotsuga macrocarpa*), and endemic populations of various cypress. species (Cupressus spp.), are especially adapted to the chaparral's natural fire regime. Coastal sagebrush (*Artemisia californica*) with other deciduous plants that tolerate drier conditions replace the evergreen chaparral at lower elevations. About 80 percent of the presettlement coastal sage scrub in southern California has been displaced, primarily by residential development.

Oak woodland is another important vegetation type in the Mediterranean California Ecoregion. Oak woodlands are also found in Arizona and New Mexico and in the Midwest. The Midwestern oak woodlands and oak savanna found in the Great Plains Ecoregion occur within a climatic zone intermediate between the prairies and Eastern temperate forests. The Central Forest/Grassland Transition Zone, consisting of a mixture of savanna, prairie, and woodlands extends from northern Illinois across much of Missouri and into eastern Kansas, Oklahoma, and Texas

(Ricketts et al. 1999). Oak savanna vegetation is now severely reduced in extent. At the time of European settlement, oak savanna was extensive, covering about 49.4 million acres (20 million ha) in a broad band stretching along the eastern edge of the Great Plains from Texas to southern Canada. There were also small areas of oak savanna found further east in Indiana, Michigan, and Ohio. Because of its open woodland nature, savannas are attractive to people and are less intimidating than dense forest. Oak savannas were among the earliest areas in the Midwest to be settled by Europeans. Trees were felled for lumber, and even when they were not cut down, their roots were frequently damaged by disturbance and branches were removed. The herbaceous layer of the savanna was cleared quickly to suppress tall "weeds." These areas were also heavily grazed. By the second half of the nineteenth century much of the savanna vegetation was destroyed. Today, midwestern savannas amount to only about 30,000 acres (12,000 ha) in total, made up of small patches, most of which are degraded.

Grasslands

The largest vegetation unit in North America is prairie, with grasses the dominant component of this vegetation. Other types of grasslands are desert grasslands of the Southwest, Californian grasslands, the palouse prairie of the Northern Rockies, the fescue (*Festuca*) prairie of northern Montana, and coastal prairie along the Gulf of Mexico. The current extent of grasslands in the U.S. is over 310 million acres (125 million ha) (Sims and Risser 2000). Grasslands depend on a climate with high light intensity, warm temperatures, and at least one dry season. The area of prairie that once extended from Canada to the Mexican border and from the foothills of the Rocky Mountains to Iowa is referred to as the Great Plains and corresponds to the Level I Ecoregion of that name. Grasslands also extended eastward into Illinois, Wisconsin and western Indiana in a region known as the Prairie Peninsula within the Level 1 Eastern Deciduous Forest Ecoregion. Climatic conditions are continental with distinct wet and dry seasons. The three main types of prairie within the Great Plains and Prairie Peninsula are tallgrass prairie, which has the highest proportion of forbs, extending from Canada and Minnesota south to Texas and east to Indiana; the mixed grass prairie, extending from Canada and eastern North Dakota south to Texas; and short grass prairie, stretching from eastern Montana south to western Texas and New Mexico. In general the prairies have few endemic species. This is often attributed to the relatively young age of these grasslands with the present distribution reflecting postglacial conditions (Sims and Risser 2000). Approximately 400 million acres (162 million ha) of prairie covered the Great Plains before European settlement (Samson et al. 2004). These prairies were maintained by frequent fires and grazing by large animals, including bison. Today relatively large portions of the short grass and some mixed grass prairie remain, although they are often degraded. The vast majority of the tallgrass prairie, however, has been converted to agricultural production, utilizing the extremely rich soils that the deep roots of prairie plants helped create. Well under 5 percent of tallgrass prairie's former extent remains today. These small and isolated prairie remnants require careful management, taking into account fire and grazing together with the need for restoration.

Deserts

The deserts of the U.S. are geographically defined by areas of low precipitation. The Cascade and the Sierra Nevada Mountains on the western boundary of the broad desert region create rainshadows and the Rocky Mountains to the east block weather fronts developing in the Great Plains. Cold deserts extending from southwestern Canada to the southern Great Basin and Colorado Plateau experience cold winters and hot summers. The Great Basin, consisting of a series of uplifted mountain ranges and intervening valleys, is the largest arid area of the U.S. The warm deserts further south, comprising the Mojave, Sonoran, and Chihuahuan Deserts have limited precipitation, mainly in the form of rain. The North American Deserts Ecoregion encompasses all these desert areas, extending from Canada to Mexico through the interior western U.S. The region differs from the Marine West Coast Forest Ecoregion to the west by its aridity and its unique shrub and cactus vegetation with a lack of trees.

The main vegetation type of the cold deserts is sagebrush steppe dominated by woody species of sagebrush. Sagebrush steppe once occupied more land than any other semidesert vegetation type in the U.S., covering about 131 million acres (53,000 ha). Now, sagebrush steppe is the dominant vegetation type of the Great Basin and Columbia Plateau. It occurs along the margins of the Mojave Desert, including in the southern slopes of the Sierra Nevada Mountains and Transverse Ranges of California, and in the Colorado Plateau and Canyonlands region. Sagebrush steppe is often associated with pinyon-juniper woodland and forests at high elevations, salt desert ecosystems at low elevations in the more arid Great Basin and Colorado Plateau, and with grasslands in the western Great Plains. Early settlers described a vast sea of sagebrush, but due to conversion of deep soil areas to cropland together with grazing, burning, and plant invasions, sagebrush communities are now estimated to exist on about 55 percent of their historical land area (Miller et al. 2011). The fragmentation of sagebrush habitats, modified and destroyed by a range of threats, has been exacerbated by the spread of invasive weeds, such as cheatgrass (*Bromus tectorum*), which fuel more intense wildfires.

Two broad groups of sagebrush species are recognized, tall and low, with different soil preferences. The most widespread big sagebrush species (*Artemisia tridentata*) currently has four recognized subspecies: basin big sagebrush (*Artemisia tridentata* ssp. *tridentata*), Wyoming big sagebrush (*Artemisia tridentata* ssp. *wyomingensis*), Mountain big sagebrush (*Artemisia tridentata* ssp. *vaseyana*), and Parish's big sagebrush (*Artemisia tridentata* ssp. *parishii*). The first three of these are dominant in the Great Basin, Columbia Basin, Snake River Plain, Colorado Plateau, and Wyoming Basin, with six other species of *Artemisia* also of importance. Parish's big sagebrush is found in coastal ranges of Southern California and south of the Great Basin.

Box 3.3: Sagebrush and Sage-Grouse Adaptive Strategies

Characteristics of sagebrush species that allow them to dominate over such large areas of the cold deserts of the U.S. include their ability to photosynthesize throughout the year due to their specialized foliage. The leaves exhibit seasonal dimorphism. Large ephemeral leaves develop in the spring and persist into the summer, whereas smaller leaves which appear in late spring last throughout the year. The smaller persistent leaves contain varying amounts of volatile chemicals, depending on the subspecies, that discourage grazing. The root system is similarly adapted to harsh climatic conditions. Big sagebrush has fibrous roots that can access water and nutrients near the surface and a taproot that can access water and nutrients at depth Welch (2005).

The greater sage-grouse (*Centrocercus urophasianus*), a species closely associated with sagebrush landscapes, is an iconic bird that has declined dramatically because of habitat loss. It is one of the more than 350 plant and animal species of conservation concern associated with sagebrush as identified by Wisdom et al. (2005). Greater sage-grouse depend almost entirely on the leaves of sagebrush for food during the winter months. Native grasses are also an important food source and provide cover more or less throughout the year. Forbs, especially in the daisy and pea families (Asteraceae and Fabaceae), are particularly important to the greater sage-grouse in spring. They provide an additional source of food, and the insects they support provide protein for hens and their young. Sagebrush acts as a "nurse" plant enabling the successful germination of other flowering plants and grasses, many of which are, in turn, important for sustaining pollinators, wildlife, and domestic livestock.

Sources: Welch (2005), Wisdom et al. (2005), Dumroese et al. (2015)

The Mojave Desert, with bi-seasonal precipitation, occurs in southeastern California and portions of Nevada, Arizona and Utah, occupying more than 25,000 square miles (65,000 square km). It is dominated by extensive stands of *Larrea tridentata* (creosote bush), and *Ambrosia dumosa* (white bursage), which cover approximately 70 percent of this desert region. Although often considered to be transitional from the Great Basin to Sonoran Desert vegetation, 25 percent of the plants of the Mojave Desert are endemic (MacMahon 2000).

The Sonoran Desert has bi-seasonal rainfall. It is considerably larger in extent than the Mojave Desert, extending from southern California through southwest Arizona and into the northwest portion of mainland Mexico and Baja California. The Californian portion of the Sonoran Desert is often known as the Colorado Desert. The Sonoran Desert has more diverse vegetation than the other U.S. warm deserts and a richer tree flora with *Parkinsonia* spp. (palo verde), *Prosopis* spp.

(mesquite) and *Olneya tesota* (desert ironwood). The shrubs, creosote bush and white bursage, cover large areas.

The Chihuahuan Desert, covering parts of Texas, New Mexico, and Arizona and an extensive area of Mexico, is at higher elevations than the Mojave and Sonoran deserts and has a cooler, wetter climate. Plants of the Chihuahuan desert scrub are often shorter with sparser foliage than similar plants of the Sonoran or Mojave deserts. Tarbush (*Flourensia cernua*) and creosote bush are dominant shrubs, and are intermixed with grasses throughout much of the Chihuahuan Desert.

Desert vegetation is uniquely adapted to withstand harsh ecological conditions. Whereas the cold deserts of the U.S. have been extensively modified, approximately half of the Mojave Desert remains as intact habitat, and the rest has not been significantly altered by human activity. About 60 percent of the Sonoran Desert has been modified and the Chihuahuan Desert is heavily degraded (Ricketts et al. 1999).

Wetlands

Wetlands include a broad range of habitats occurring throughout the ecoregions of the U.S. and in every state where appropriate conditions prevail. They can be defined as "*lands transitional between terrestrial and aquatic systems where the water is usually at or near the surface or the land is covered by shallow water*" (Cowardin et al. 1979). Wetlands include coastal habitats such as salt marshes and freshwater habitats associated with lakes, rivers and peatlands. In the Northern Forests Ecoregion, peatlands are a characteristic feature, for example in northern Minnesota. Marshes and fens known as prairie potholes, occur in northern Montana, North and South Dakota, Minnesota, and Iowa. Prairie potholes are extremely important for nesting waterfowl.

A nationwide inventory of U.S. wetlands is maintained by the U.S. Fish and Wildlife Service. The National Wetlands Inventory (NWI), is based on a wetland classification system first published by Cowardin et al. (1979), which has become a Federal standard. Mapping and recording is based on aerial imagery. In 2009 the FWS recorded an estimated 110.1 million acres (44.6 million ha) of wetlands in the conterminous U.S. An estimated 95 percent of all wetlands were freshwater and 5 percent were in marine or estuarine (saltwater) systems. Salt marsh made up an estimated 66.7 percent of all estuarine and marine wetland areas. Forested wetlands made up the single largest category (49.5 percent) of freshwater wetlands with freshwater emergents making up around 26 percent, shrub wetlands 17.8 percent and freshwater ponds 6.4 percent by area.

The Everglades is perhaps the most celebrated wetland area of the U.S. It is unique among the world's largest wetlands in that its water is derived from rainfall (Rickets et al. 1999). This complex and huge wetland ecosystem covers 2 million acres (0.8 million ha), extending from central Florida, near Orlando, south to Florida Bay. Lake Okeechobee, the seventh largest lake in the U.S., is the source of much of the water. During the wet season the large lake overflows, releasing its water into a very slow moving river dominated by sawgrass marsh. The water flows southward, passing through diverse habitats, including cypress swamps, wet prairie, and mangroves, until it reaches Everglades National Park and eventually Florida Bay. The Everglades has extensive areas of peatland. This massive wetland area has been extensively modified through drainage, agriculture, and urban development for the past 200 years. The Everglades is a wetland site of international interest as recognized by the Ramsar Convention. A list of other internationally important wetland sites in the U.S. is given in Appendix 2. Wetlands throughout the U.S. are of immense ecological importance in the maintenance of water supplies, for carbon storage, and in the provision of wildlife habitat. Coastal wetlands are particularly important in buffering the coastline against storm surges. This is particularly important given that U.S. coastal counties now account for nearly 40 percent of the total U.S. population.

Box 3.4: Ecosystems and Carbon Storage

Terrestrial and aquatic ecosystems play an important role in the carbon cycle. Major ecosystems that commonly are considered in both global and national carbon assessments and inventories include forests, croplands, grassland/shrublands, and wetlands. Plant biomass, both above and below ground, is a major pool of carbon with photosynthesis being the driving process behind carbon storage in plant material. The amount of carbon stored in plant biomass is influenced by land use. For example, forest clearing for cropland greatly reduces the amount of carbon stored in the vegetative biomass. In a natural system, most of the biomass production contained in living plant material is eventually transferred to dead organic matter such as dead wood and litter. Dead organic matter on the ground and plant biomass below the ground decompose and transform into soil organic matter, which is another primary carbon pool. Decomposition of soil organic matter releases carbon dioxide back into the atmosphere.

Recent studies indicate that terrestrial ecosystems in the U.S. represent a sizeable and globally important carbon sink (Potter et al. 2007). A methodology for assessing carbon stocks in U.S. ecosystems is presented by Zhu et al. (2010). In this methodology, when plant species are evaluated as part of assessing management activities, only native plant species are considered. Zhu et al. (2010) report that forests are a large carbon sink, but they are ecosystems that gain and lose carbon continually. The overall carbon stock of forests in the U.S. is estimated as 57,000 TgC (trillion grams of Carbon). The equivalent figure for grassland and shrublands combined is 16,000 TgC with most of the carbon stock stored in the soil. For wetlands the estimated carbon storage is 62,000 TgC.

Source: Potter et al. (2007), Zhu et al. (2010)

Changing Ecosystems

Ecosystems are dynamic with constant interaction between biotic and abiotic factors. All ecosystems change and adapt over time and at different timescales. All ecosystems in the U.S. have been radically altered since the time of European settlement. The pace of environmental change globally has increased through the so-called Anthropocene Era during which humans and our societies have become a global geophysical force. It is generally recognized that the Anthropocene began around the year 1800 at the onset of industrialization in Europe and in the eastern U.S. and that since 1945 the world has witnessed the second stage, the so-called Great Acceleration. Species extinctions and climate change are two of the interlinked outcomes. There is increasing urgency to understand and address ecosystem loss and modification throughout the ecoregions of the U.S. Management of land now requires incorporating climate change mitigation and adaptation measures. In Chap. 5 we discuss the history of land use in the U.S., which agencies are currently responsible for managing and restoring the land, and their roles in relation to climate change. In Chap. 6 we discuss the procedures and initiatives for conserving and restoring ecosystems and the diversity of plant species within them.

References

Barbour MG, Billings WD (eds) (2000) North American terrestrial vegetation, 2nd edn. Cambridge: Cambridge University Press, 708 p

Brockway DJ, Outcalt KW, Tomczak DJ, Johnson EE (2006) Restoration of longleaf pine ecosystems. USDA Forest Service General Technical Report SRS-83, Asheville, NC, 34 p

Commission for Environmental Cooperation (1997) Ecological regions of North America. Toward a common perspective. Commission for Environmental Cooperation, Montreal, Quebec, 71 p

Cowardin LM, Carter V, Golet FC, LaRoe ET (1979) Classification of wetlands and deepwater habitats of the United States. U.S. Fish and Wildlife Service, FWS/OBS-79/31, Washington, D.C.

Delcourt HR, Delcourt PA (2000) Eastern deciduous forests. In: Barbour MG, Billings WD (eds) North American terrestrial vegetation, 2nd edn. Cambridge University Press, Cambridge, pp 357–395

Dumroese RK, Luna T, Richardson BA, Kilkenny FF, Runyon JB (2015) Conserving and restoring habitat for greater sage-grouse and other sagebrush-obligate wildlife: the crucial link of forbs and sagebrush diversity. Native Plants J 16(3):276–299

Federal Geographic Data Committe (2008) National Vegetation Classification Standard, Version 2. FGDC-STD-005-2008 (Version 2). 126p https://www.fgdc.gov/standards/projects/vegetation/NVCS_V2_FINAL_2008-02.pdf

Gori D, Bate J (2007) Historical range of variation and state and transition modeling of historical and current landscape conditions for pinyon-juniper of the Southwestern U.S. Prepared for the USDA Forest Service, Southwestern Region by The Nature Conservancy, Tucson, AZ, 141 p

MacMahon JA (2000) Warm deserts. In: Barbour MG, Billings WD (eds) North American terrestrial vegetation, 2nd edn. Cambridge University Press, Cambridge, pp 285–322

Miller RF et al. (2011) Characteristics of sagebrush habitats and limitations to long-term conservation. USGS Administrative Report in Knick ST, Connelly JW (eds) Greater Sage-Grouse: ecology of a landscape species and its habitats. University of California Press, Berkeley, CA, Cooper Ornithological Union, pp 145–184

Omernik JM (1987) Ecoregions of the conterminous United States. Map (scale 1:7,500,000). Ann Assoc Am Geogr 77(1):118–125

Oswalt SN, Smith BW, Miles PD, Pugh SA (2014) Forest resources of the United States, 2012: a technical document supporting the Forest Service update of the 2010 RPA assessment. US Forest Service General Technical Report WO-91. U.S. Department of Agriculture, Forest Service, Washington, D.C., 218 p

Peet RK (2000) Forests and meadows of the Rocky Mountains. In: Barbour MG, Billings WD (eds) North American terrestrial vegetation, 2nd edn. Cambridge University Press, Cambridge, pp 75–122

Potter C, Klooster S, Hiatt S, Fladeland M, Genovese V, Gross P (2007) Satellite-derived estimates of potential carbon sequestration through afforestation of agricultural lands in the United States. Clim Chang 80(3–4):323–336. https://doi.org/10.1007/s10584-006-9109-3

Ricketts TH, Dinertein E, Olson DM et al. (1999) Terrestrial ecoregions of North America: a conservation assessment. Island Press, Washington, D.C., 485 p

Samson FB, Knopf FL, Ostli W (2004) Great Plains ecosystems: past, present, and future. USGS Staff -Published Research Paper 45. Wildlife Soc Bull 32:6–15

Schwartzberg L (2014) Viewing nature's beauty through a new lens. https://blog.nationalgeographic.org/2014/04/05/viewing-natures-beauty-through-a-new-lens/

Sims PL, Risser PG (2000) Grasslands. In: Barbour MG, Billings WD (eds) North American terrestrial vegetation, 2nd edn. Cambridge University Press, Cambridge, UK, pp 323–356

Welch BL (2005) Big sagebrush: a sea fragmented into lakes, ponds, and puddles. RMRS-GTR-144. USDA Forest Service, Rocky Mountain Research Station, Fort Collins, 210 p

Wisdom M Jr, Rowland MM, Suring LH, Schueck L, Meinke CW, Knick ST (2005) Ch 1. Evaluating species of conservation concern at regional scales. In: Wisdom MJ, Rowland MM, Suring LH (eds) Habitat threats in the sagebrush ecosystem – methods of regional assessment and application in the Great Basin. Alliance Communications Group, Lawrence, KS, pp 5–75

Zhu Z (ed), Bergamaschi B, Bernknopf R, Clow D, Dye D, Faulkner S, Forney W, Gleason R, Hawbaker T, Liu J, Liu S, Prisley S, Reed B, Reeves M, Rollins M, Sleeter B, Sohl T, Stackpoole S, Stehman S, Striegl R, Wein A (2010) A method for assessing carbon stocks, carbon sequestration, and greenhouse-gas fluxes in ecosystems of the United States under present conditions and future scenarios: U.S. Geological Survey Scientific Investigation Report 2010, 1144, 195 p. Available at http://pubs.usgs.gov/sir/2010/5233/

Chapter 4
Conservation of Plant Species

If we of this generation destroy the resources from which our children would otherwise derive their livelihood, we reduce the capacity of our land to support a population, and so either degrade the standard of living or deprive the coming generations of their right to life on this continent. (Theodore Roosevelt 1909)

Native plants are a natural resource of immense value. Maintaining the diversity of native plant species is fundamentally important for human well-being. In the U.S., native plant diversity is relatively well documented compared with many other parts of the world. However, its role in provisioning ecosystem goods and services is less well understood. Unfortunately, native plant species are far from secure with approximately one-third of the native flora threatened with extinction. This represents a potential major loss of species and genetic diversity, which would exert a major impact on associated insect, bird, and other animal species. The loss of individual species may potentially trigger broader ecological collapse. Furthermore, the loss of plant diversity limits our ability to adapt to future needs at a time of rapid global change. The financial and other livelihood implications of natural resource loss anticipated by Roosevelt (1909) are being increasingly documented and have profound economic implications.

Many different governmental and non-governmental organizations and individuals are devoting their time to tackling the threats to biodiversity and to safeguarding plant diversity, and these efforts need to be scaled up. The protection of natural areas and threatened species must be supplemented by restoration action on a landscape scale to reverse the tide of biodiversity loss. Mitigation as a concept has been less generally applied, although it has been inherent in environmental regulation since the Fish and Wildlife Coordination Act of 1934 (see Appendix 1). The term essentially means to avoid, minimize, and compensate for adverse impacts. Mitigation measures may be required in relation to securing endangered species and conserving and repairing damaged ecosystems. The term has come to broader prominence in relation to tackling climate change.

S. F. Oldfield et al., *Seeds of Restoration Success*, Springer Earth System Sciences, https://doi.org/10.1007/978-3-319-96974-9_4

Box 4.1: Extinct Plant Species of the U.S.

An early U.S. plant extinction in the wild was the Franklin tree (*Franklinia alatamaha*). An attractive, woody plant with fragrant, cream-flowers, this species was discovered by plantsman John Bartram and his son William in 1765 and named after Dr. Benjamin Franklin. It grew in acidic, boggy soils along the banks of the Altamaha River in Georgia. Naturally rare, overcollection by the horticulture industry subsequently contributed to its extinction in the wild.

Currently, NatureServe records 154 plant species of the U.S. as being extinct or possibly extinct in the wild. These include: bigleaf scurfpea (*Orbexilum macrophyllum*), formerly known from Indiana and Kentucky and banded Trinity (*Thismia americana*), which was last seen in Illinois in 1916 and was declared extinct in 1995. Banded Trinity was an extraordinary parasitic plant that grew in an area of wet prairie in Chicago.

In Hawaii, over 100 plant taxa (species, subspecies and varieties) are extinct and twice this number have less than 50 individuals remaining in the wild. Continued monitoring of wild populations is constantly required together with action to prevent further extinctions in the state.

It can be difficult to determine if a plant species is actually extinct in the wild without repeated searches of its known and potential habitat. The IUCN considers a species to be extinct when there is no reasonable doubt that the last individual has died. The California Native Plant Society runs the De-extinction Project, which is re-evaluating the status of 22 plant species thought to be extinct in the state. These species have not been seen for many years or occurred on habitat that was significantly degraded or destroyed. In 2016, one such species, the serpentine canyon monkeyflower (*Erythranthe pericaulis*), was rediscovered in the Plumas National Forest in the northern Sierra Nevada.

Sources: CNPS (2017), NatureServe (2019), State of Hawaii (2019)

Conservation Assessment, Planning, and Action

To prevent the loss of native plant diversity it is of fundamental importance to understand the distribution and relative abundance of species and the threats they face. Globally, the IUCN provides the most comprehensive assessment of the conservation status of species and in the U.S. the most comprehensive source of information is provided by NatureServe. This non-profit organization maintains an online database recording the status of biodiversity in the U.S., Latin America, and Canada. Each native species is assigned a conservation status rank from 1 to 5 at the global (G rank), national and state level. The conservation ranks are based on a series of criteria including size of distribution range, species population trends, and threats. Currently, NatureServe identifies about one-third (5,935 species) of the U.S. native flora as extinct, possibly extinct, or threatened in the wild. Information on conservation status is essential as a baseline for monitoring future impacts, deciding priorities, and planning conservation action. Conservation assessments of individual species can also act as indicators of biodiversity status more broadly.

Box 4.2: The NatureServe Network

The NatureServe Network, also known as The Natural Heritage Network, is a public private partnership that provides integrated, comprehensive, authoritative data on rare plants, animals, and ecosystems.

The NatureServe Network is comprised of over 80 U.S. Natural Heritage Programs and Canadian Conservation Data Centres. Many U.S. Natural Heritage Programs are affiliated with state government agencies, such as Departments of Natural Resources, while others are parts of universities or non-governmental organizations. For more information, visit natureserve.org.

The NatureServe Network, with over 1,000 scientists, collects and manages data on the conservation status and distribution of species and ecosystems of conservation concern. Scientists use a standardized methodology, data structure, and database, allowing information to be integrated and rolled-up into cohesive national and international datasets. NatureServe's Conservation Status Assessments or Ranks are used extensively in the U.S. and Canada to prioritize conservation action. For ecosystems and natural communities, databases contain information on vegetation structure, composition, succession patterns, natural disturbances, conservation status, and distribution of specific ecological community types throughout their geographic range. Data from the NatureServe Network has many applications, from land use and management decisions to describing reference communities for restoration. Information on ecological communities and species can be found on NatureServe Explorer (explorer.natureserve.org).

Sources: NatureServe (2019)

Conservation assessments have been assigned in the U.S. to give legal protection to plant species. The Endangered Species Act (ESA) was signed into law in 1973. It aims to protect species from extinction by prohibiting actions that threaten their survival. The protected species are listed as Threatened or Endangered following a lengthy process of scientific and legal review. The criteria for listing are: (a) present or threatened destruction, modification or curtailment of its habitat or range; (b) overutilization for commercial, recreational, scientific, or education purposes; (c) disease or predation; (d) the inadequacy of existing regulatory mechanisms; and (e) other natural or manmade factors affecting its continued existence. At present 944 plant taxa are listed as Threatened or Endangered under the ESA, a relatively small proportion of the threatened plants identified by NatureServe.

Assessment of conservation status now requires consideration of climate change. Guidelines are provided for assessing IUCN threat status categories in relation to the changing climate. In general, species distribution models are used to predict the impact of climate change. For each species a map of present distribution determined by climatic conditions provides a "climate envelope." Predicted temperature and rainfall attributes are then mapped to see how future climate might shift the species distribution. A 1996 study indicated that with a mean global warming of 3 °C, about 7–11 percent of North American plant species would be entirely out of their climate envelopes and thus vulnerable to extinction if they are unable to adapt or migrate to track climate. Species already rare and under threat (those with NatureServe global ranks 1-3) would be disproportionally affected with 10–18 percent out of their climate envelopes (Kutner and Morse 1996).

Recent studies have looked at the impact of climate change on plant diversity in the western U.S. They have shown that patterns of future plant loss may not be correlated with current conservation status, such as NatureServe's global rank, federal Endangered Species Act status, or state level rare plant ranks (Still et al. 2015), indicating that these systems may no longer be sufficient to prioritize conservation.

Still et al. (2016) modeled the future distributions of 565 western rare plant species and showed that 45 percent are predicted to be at the highest risk of extinction by the 2080s. Of the 565 taxa modeled, 16.8 percent are listed as either federally Endangered or Threatened. Of these, 44.2 percent are predicted to be losing suitable habitat by the 2080s, as compared to 55.8 percent of the taxa that are not listed. This indicates that federally listed taxa are not more likely to be at risk because of climate change than non-listed species. In the same way, a higher rarity status as indicated by NatureServe global rank does not indicate increased vulnerability to climate change with 44.0 percent of the G1 species predicted to be losing suitable habitat by the 2080s, as compared to 53.8 percent of the G2 taxa and 41.2 percent of the G3 taxa. The authors summarized that the majority of the rare plants in the western U.S. are predicted to be threatened by climate change. However, rarity was not necessarily correlated to climate change vulnerability.

Following species conservation assessment, action is required for species facing extinction. Research to understand the causes of decline and develop appropriate responses is often a key component of recovery planning. As far as possible the threatening factors should be controlled or removed. Increasing awareness and influencing behavior are generally very important species conservation measures. Two broad forms of species conservation are widely recognized, with conservation of the species in its natural habitat or in situ conservation usually considered the preferred option. Ex situ conservation is the conservation of a species outside its habitat. Ex situ plant material can be used for research, propagation, and ideally reintroduction to enhance wild populations.

In Situ Conservation

Some threatened plants of the U.S. are specifically protected in their natural habitats by federal and state legislation and policy. Others may be given a degree of protection by legal and policy provisions relating to the conservation of specific habitats. Once a species is listed as Threatened or Endangered under the Endangered Species

Act (ESA), the U.S. Fish and Wildlife Service is required to create a recovery plan to improve management of the species and increase its numbers so that it has a greater chance of survival. Recovery plans have not yet been produced for all federally listed plant species, and in some cases further research is needed to determine which conservation measures will be effective. It has been suggested that recovery planning may be further complicated in the era of climate change as *"there may be few practical domestic mechanisms to promote recovery in the face of a global problem"* (Liebesman et al. 2009).

Furthermore, plants listed under the ESA are only protected from damaging activities if they occur on federal lands. This is not the case for listed animal species that are protected wherever they occur. This anomaly persists because federal and state wildlife laws in the U.S. have their origins in old English common law where the King and Parliament owned the wild animals and prescribed the ways that ordinary citizens could harvest them on both public and private lands. Conversely, plants were considered to be a part of the land on which they grew, and thus plants on private lands were and are treated as the owner's private property (Havens et al. 2014).

In addition to the ESA Federal Act, many states have their own legislation protecting plants. In 15 states, plants are part of the overall state endangered species acts. In 17 other states, plants and animals have separate protective acts. Most state laws provide fewer and weaker protections for endangered plants than the ESA, usually with minimal penalties for infringement, but they nevertheless provide a degree of protection. Like the ESA, most state laws provide no or minimal protection for endangered plants on private lands; therefore, involving landowners in plant conservation solutions is essential. Very few state laws require state agencies to consult on their projects that may damage endangered plants. In addition, few have provisions for the designation of critical habitat or a requirement to develop recovery plans. Nevertheless, all state laws designate an agency responsible for endangered plants and require the agency to develop a state list. The state agency is usually responsible for informing and involving other agencies, organizations, and individuals in conserving the species.

As noted above, protecting threatened plant species in situ in the natural habitats where they grow is generally recognized as the best conservation option wherever this is practicable. The potential for and reality of in situ plant conservation depends on land ownership, and this is why working with all landowners is so important. As discussed in Chap. 5, federal lands, which make up approximately one-third of the U.S., are generally managed for multiple and at times conflicting land uses. Areas specifically set aside for conservation include Bureau of Land Management's National Conservation Lands, National Parks and other areas managed by the National Park Service, and the interagency National Wilderness Preservation System, which includes more than 107 million acres (43 million ha) in 44 states. In total, about 27 percent of the land of the U.S. is under some form of conservation management, and a map of major protected areas is shown in Fig. 4.1. In 2008, there were 831 plants protected under the ESA on federal lands in the United States. An additional 2,686 plant species on federal lands were considered imperiled by NatureServe (Stein and Gravuer 2008). In 2019, 944 plants are listed under the ESA, comprising 56 percent of all U.S. species protected under the ESA.

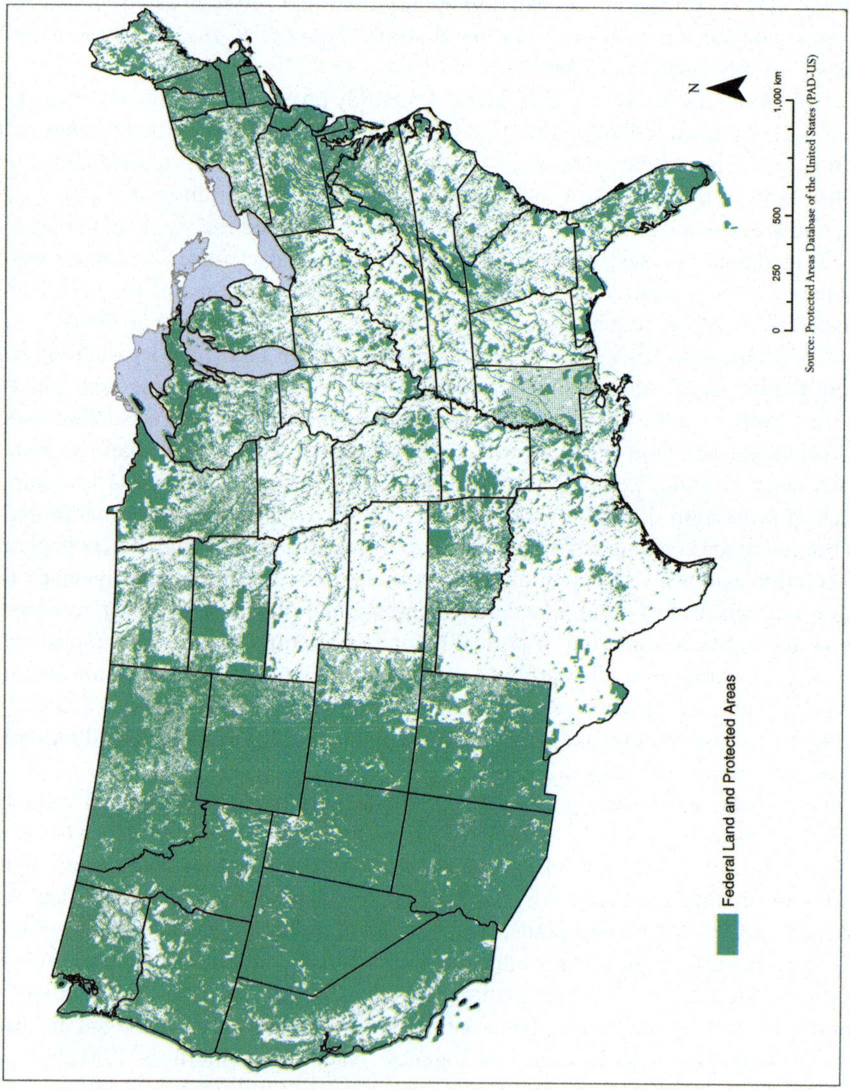

Fig. 4.1 Protected areas in the United States including public land and voluntarily provided private protected areas (USGS 2018). Major protected areas of the U.S.

Box 4.3: The Gap Analysis Program (GAP)

Regional, state, tribal, and federal agencies, academic institutions, private companies, and conservation groups—a total of some 500 organizations—participate in the Gap Analysis Program, a nationwide initiative coordinated by the U.S. Geological Survey (USGS) to assess the extent native animal and plant species are being protected. The national guidelines define three designations for protected areas

GAP Class 1: An area having permanent protection from conversion of natural land cover and a mandated management plan in operation to maintain a natural state within which disturbance events (of natural type, frequency, intensity, and legacy) are allowed to proceed without interference or are mimicked through management. This class includes federal designations such as national parks, national monuments, wilderness areas, nature reserves, preserves, and research natural areas.

GAP Class 2: An area having permanent protection from conversion of natural land cover and a mandated management plan in operation to maintain a primarily natural state, but which may receive uses or management practices that degrade the quality of existing natural communities, including suppression of natural disturbance. This class includes state parks, state recreation areas, national wildlife refuges, national recreation areas, areas of critical environmental concern, wilderness study areas, conservation easements, private conservation lands, and national seashores.

GAP Class 3: An area having permanent protection from conversion of natural land cover for the majority of the area, but subject to extractive uses of either a broad, low-intensity type (e.g., logging) or a localized intense type (e.g., mining). The designation also confers protection to federally listed endangered and threatened species throughout the area. This class includes national forests, most BLM land, wildlife management areas, military reservations, state forests, game and fish preserves, state commemorative areas, access areas, national grasslands, and Army Corps of Engineers holdings.

The most recent analysis shows that more than 753 million acres (305 million ha) of land is classified under one of the three protection categories.

Source: USGS (2019)

State and local governments also maintain protected areas set aside for biodiversity conservation. In addition, non-governmental organizations (NGOs) also manage significant areas of land for conservation purposes. The California Native Plant Society (CNPS), with its 10,000 or so members, has a renowned program for the preservation of native plant species and their habitats on both public and private lands. Advocacy for the maximum protection of native plants is an important part of the work of CNPS, together with the promotion of science-based and ecologically sound land management practices.

California has two of the oldest national parks established in the U.S., Yosemite National Park and Sequoia National Park, created within days of each other in 1890. Other important national parks in the region include Redwood National Park, officially established in 1968 (and expanded in 1978) and the 390 square mile (1,010 km^2) Channel Islands National Park, a series of islands off the coast of southern California. Nearly 50 wilderness areas in California are managed by the USFS, as well as 16 national wildlife refuges managed by FWS, and six military installations. In addition, BLM manages more than 15 million acres (6 million ha) in California, of which 65 percent is National Conservation Lands. There are also 107 state parks and more than 50 areas managed by NGOs. The creation of many of these protected areas was the result of the dedicated efforts of national conservation organizations, including the Sierra Club, The Nature Conservancy, and The Wilderness Society.

Relatively few protected areas in the U.S. have been designated primarily to conserve iconic or threatened plants or plant diversity, but there are exceptions. Research Natural Areas (RNAs) are areas designated by the USFS to be permanently protected and maintained in natural condition. RNAs include unique ecosystems or ecological features; rare or sensitive species of plants and animals and their habitat; or high-quality examples of widespread ecosystems. Furthermore, some national designations protect iconic species. In 1933, Saguaro National Monument was created in the Coronado National Forest, Arizona, because "*the exceptional growth thereon of various species of cacti, including the so-called giant cactus, it appears that the public interest will be promoted by reserving as much land as may be necessary for the proper protection thereof*" (Hoover, Presidential Proclamation 1 March 1933). The Saguaro Wilderness Area of 71,400 acres (28,900 ha) was added in 1975. Saguaro National Park was created from these areas in 1994 and currently encompasses 91,327 acres (36,959 ha) in its two districts. Another site protected for its botanical heritage is the Organ Pipe National Monument, created in 1937 as a way to preserve a representative area of the Sonoran Desert, with its 26 species of cacti and stunning spring wildflowers. Covering more than 500 square miles (1300 km^2) and lying on the border of the U.S. and Mexico, the area, which was designated as an internationally recognized Biosphere Reserve in 1976, protects a wilderness area of mountains, rocky canyons, and sweeping plains.

In California, Redwood National Park protects the habitat of some of the world's tallest trees: old-growth coast redwoods. These magnificent trees can live to be 2,000 years old and grow to over 300 feet (91 m) tall. Spruce, hemlock, Douglas fir; shrubs such as rhododendron (*Rhododendron macrophyllum*), huckleberry, salal (*Gaultheria shallon*), together with sword ferns (*Polystichum*) create a multiple-canopied understory. Redwood National Park's mosaic of habitats includes prairie and oak woodlands and 37 miles (59.5 km) of Pacific Ocean coastline.

Usually plants within National Parks and other protected areas are conserved by their occurrence in sites designated for nature conservation. To be truly effective, conservation may require additional forms of management. This may include removal of invasive species, prescribed fire, altering grazing regimes, prevention of trampling, and enhancing natural populations. At the very least, monitoring of rare plant populations is required. In some states, programs have been established to monitor rare plants in their natural habitats. The Colorado Natural Heritage Program (which tracks more than 520 plant species in Colorado, including the number and mapped locations of populations, number of individuals, and threat and quality assessments) has established Important Plant Areas throughout Colorado. They are summarized in the Colorado Rare Plant Strategy, developed by the Colorado Rare Plant Conservation Initiative. Rare plant monitoring is also conducted by NGOs, often in partnership with citizen scientists. For example, the Chicago Botanic Garden's Plants of Concern Program monitors nearly 300 rare plant species with the assistance of approximately 200 dedicated volunteers in Illinois and Indiana. The monitoring data are shared with Illinois and Indiana's Natural Heritage programs and inform management.

Box 4.4: The Importance of Understanding Plant Conservation Genetics
Genetic diversity is what allows species to adapt to changing climates and other new threats and challenges. Therefore, maintaining adaptive genetic diversity, whether in populations in situ or ex situ, is one of the primary goals of conservation programs. Genetic diversity, both in wild and in ex situ populations is important for many reasons. Genetically diverse populations typically have broader ecological tolerances and can better survive challenges, such as disease and weather extremes. Diverse populations have higher fitness and less inbreeding depression. And diversity increases mating opportunities in self-incompatible species.

Conservation genetics provides tools to help guide conservation and restoration efforts, to measure and monitor success, and ultimately to minimize the risk of extinction for all plant species by preserving them as dynamic entities capable of evolving in response to changing environments.

Genetic studies have been used to assess genetic variation and develop conservation plans. These typically fall into two broad categories quantitative genetic studies and molecular genetic studies. Quantitative studies can gauge the ability of a population to adapt because they look at the genetic diversity upon which selection acts, a critical determinant of a population's evolutionary potential. Molecular studies look at neutral markers like allozymes and microsatellites, which can help resolve taxonomic questions, detect hybridization and inbreeding, and assess gene flow within and between populations. Both types of studies can provide valuable information for conservation planning.

Genetic issues and questions can come into play in all types of conservation activities, including:

- assessing population viability
- collecting genetically adequate samples for seed banking or breeding programs
- determining conservation units for management
- maintaining diversity in *ex situ* collections
- determining genetically-appropriate seed mixes for reintroduction

Source: Kramer and Havens (2009)

A new challenge in plant conservation is to anticipate future climatic conditions that are fundamentally changing the distribution of plants and their natural habitats. Specific adaptation strategies include conserving species populations with greater genetic diversity, managing relocation in recovery planning (known as assisted migration), and boosting ex situ conservation collections.

Recovery plans for individual plant species are hugely important to reverse the tide of extinction. Recovery may include reinforcement of populations in the wild or reintroduction of plant populations to land where they previously occurred. One quarter of recovery plans for plant species listed under the ESA include a requirement for reintroduction. At the same time, much broader work is increasingly required to recover endangered or degraded habitats so that rare, threatened, and more common plant species can all flourish. With the impacts of climate change, it is increasingly important to broaden the emphasis to include all native species in conservation planning. Ecological restoration aims to restore natural resources and ecological functioning at the ecosystem level. Whatever the scale of ecological restoration, availability of native plant materials for both rare and common species is essential for the recovery process. Exciting and ambitious schemes to develop plant materials for restoration of land at the regional level with a focus on the Great Basin, Colorado Plateau, and Mojave Desert of the North American Deserts Ecoregion; the Pacific Northwest; east coast wetland habitats; and at a national level are discussed further in Chap. 5.

Ex Situ Conservation

Plant rich habitats throughout the U.S. deserve and urgently need conservation attention to ensure their survival. Restoration of degraded lands is also vitally important to guarantee the continued ecological health of the nation. Many dedicated organizations and individuals are working tirelessly to conserve and restore native plant communities. Complimentary to these efforts in conserving plants in the wild is ex situ conservation, defined as the conservation of plants and animals outside their natural habitats. Botanic gardens, arboreta, and other living botanical collections, such as seed banks are helping to prevent species extinctions by strategically building and using their collections to support threatened species conservation. As the threats to individual species and their broader assemblages intensify, it is becoming increasingly important to build up ex situ stocks of all native plants to keep future options open.

High quality ex situ collections provide a valuable source of plant material, which can be used for reintroduction of threatened species to the wild and for the restoration of damaged or degraded habitats. It is critical that plant material used for reintroduction is carefully matched to the habitat, and the genetics of the local, wild populations. Plants in living collections can also be used for research on endangered plants, which is often necessary for successful species recovery programs.

Over the past 20 years, Seeds of Success (SOS) has led the way nationally in collecting native seed from the wild. Established by BLM as part of the BLM Plant Conservation and Restoration Program, the SOS teams have made more than 24,000 native seed collections covering more than 5,500 taxa for use in restoration, ex situ conservation, and in developing native seed crops.

For some plants, where natural habitat is no longer available, ex situ conservation may, unfortunately, be the only option in the fight against extinction. Botanic gardens play important roles in both in situ and ex situ plant conservation, but they are more usually associated with ex situ conservation of wild plants because they are the main institutions equipped to do this. The U.S. has over 400 botanic gardens and arboreta, some of which manage natural areas for conservation (Fig. 4.2). Behind the scenes of the horticultural display areas of many gardens are herbaria, laboratories, glasshouses, field nurseries, and seed banks dedicated to the ex situ conservation of imperiled plants and their recovery.

Progress in ex situ conservation of imperiled species in the U.S. is encouraging. One of the defining features of botanic gardens is that they document their collections. It is therefore possible to monitor which species are being grown by botanic gardens or stored in seed banks. According to the PlantSearch database maintained by Botanic Gardens Conservation International (BGCI), 39 percent of the U.S. threatened plants listed by NatureServe are currently conserved in ex situ holdings (Kramer et al. 2011). Furthermore, the U.S. has nearly three-quarters of its federally listed taxa in ex situ conservation holdings, and 65 percent of these taxa are found in genetically diverse germplasm collections.

Ex situ conservation of federally listed Threatened and Endangered plant species has been coordinated for more than 35 years by the Center for Plant Conservation (CPC). Acting in this way and through advocacy CPC is a force for plant conservation at the national level. The aim of the organization is to conserve (in seed banks and living collections) and restore the rare native plants of the U.S. CPC has oversight responsibility for the National Collection of Endangered Plants, which contains plant material for over 1,400 of the country's most threatened species. The Center operates through a cooperative network of 43 participating institutions located throughout the country from Hawaii to Massachusetts. Each member botanic garden takes responsibility for looking after one or more threatened plants. The Lady Bird Johnson Wildflower Center in Austin, Texas, for example, works on the ex situ conservation of 14 globally rare taxa for CPC, including Texas wild rice (*Zizania texana*) and Pecos sunflower (*Helianthus paradoxus*), two wild relatives of crop species referred to in Chap. 2.

Another example of a threatened taxon in the National Collection of Endangered Plants is Texas snowbells (*Styrax platanifolius* ssp. *texanus* previously known as *S. texana*), another of the species for which Lady Bird Johnson Wildflower Center acts as primary custodian. Texas snowbells is an attractive, deciduous tree with white bell-shaped flowers and heart-shaped leaves. One of the special plants of the Edwards Plateau, Texas snowbells, is federally listed as Endangered and its recovery plan was produced in 1987. Subsequent detailed research has located more individuals in the

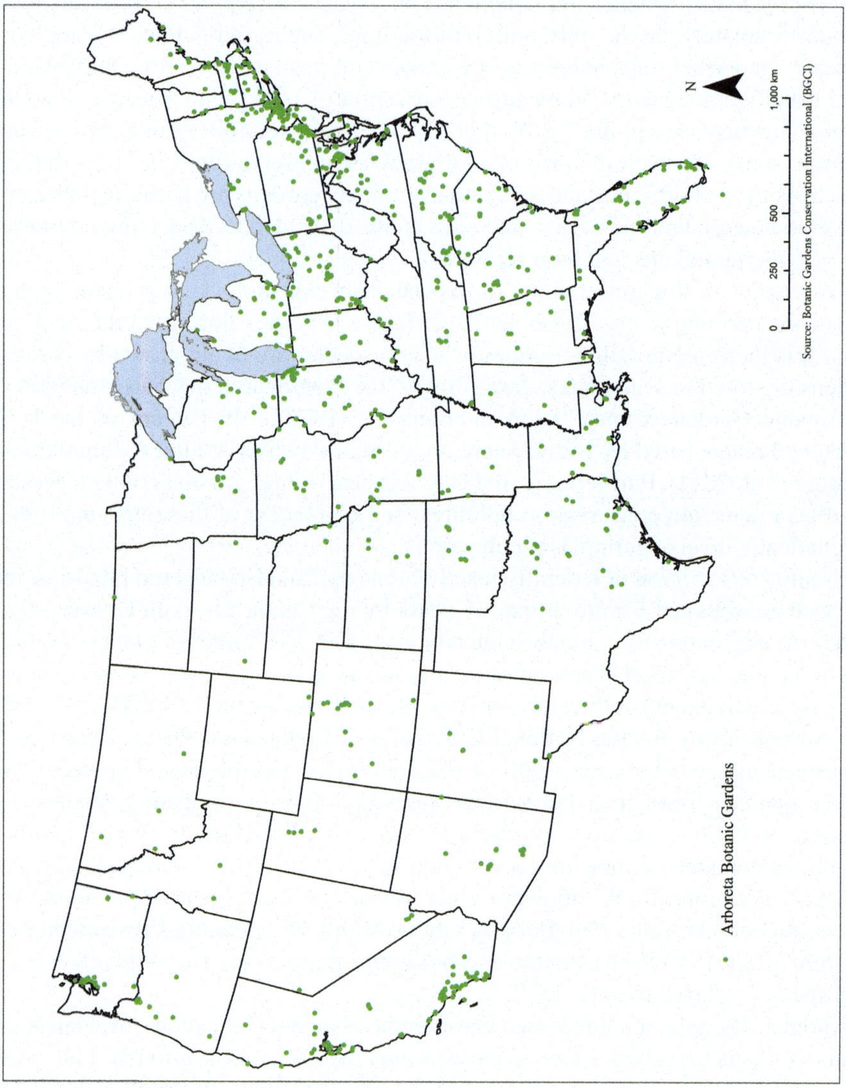

Fig. 4.2 Botanic gardens and arboreta in the U.S. (BGCI 2019)

wild, but only about 1,000 are known to remain. Regeneration of this species has been a problem due, in part, to heavy grazing by domestic livestock and exotic game. The few remaining mature trees grow close to streams, and most seeds are carried away by floodwater when they fall from the plant. The seedlings that do survive are frequently eaten by deer, wild goats, sheep, or rodents.

Chicago Botanic Garden is a CPC member working in the upper Midwest. One of the plants in its care is Pitcher's thistle (*Cirsium pitcheri*). Ex situ conservation, primarily through seed banking, is especially important for this species, which is being driven toward extinction in the wild by several species of weevil used for biocontrol of weedy thistles. This beautiful native thistle grows only on the sand dunes of the western Great Lakes where development, recreation, and climate change have also taken their toll on wild populations. Because the dunes are fairly species poor, Pitcher's thistle plays a critical role in supporting over 50 pollinator species when it is in bloom (Havens and Vitt, unpub. data). Its extinction could cause significant pollinator loss, which in turn could affect other plants that call the dunes home.

Another example of a plant conserved by CPC is Stern's medlar (*Mespilus canescens*), one of the endangered (G1) plant species that Missouri Botanical Garden is helping to conserve. This tree grows only in one small area of the U.S., and its only close relative is the common cultivated medlar, the fruit tree which occurs naturally in Southeast Europe and West Asia. Stern's medlar was not discovered until 1990. The only known wild trees occur in the Mississippi Delta region of Arkansas, where a population of around 25 individuals clings to survival. The wild trees produce flowers, but not fruit or seed. Most of the natural woodland habitats in the area have been cleared for farming. Fortunately, a private landowner has taken steps to ensure that Stern's medlar is protected through an agreement with the Arkansas Natural Heritage Commission. The wild trees are carefully monitored to make sure that they are not damaged by agricultural activities. The long term chances of survival have been enhanced by establishing rooted cuttings at Missouri Botanical Garden for research purposes, which might ultimately lead to the re-establishment of wild populations.

CPC members work with threatened plants both in ex situ collections and in the wild. In the greenhouse, scientists conduct horticultural research and establish how to grow and propagate the threatened plants. These scientists provide plant material for restoration efforts, and they also assist in monitoring populations in the wild, managing habitats, and restoring plants to native habitats. CPC institutions are currently involved in more than 80 restoration projects. An important activity being undertaken is to study the status of all plants listed under the ESA and look at their potential for recovery. CPC's work is carried out in close collaboration with landowners and land managers, ensuring an integrated approach to plant conservation. The conservation work of CPC is a model for plant conservation efforts around the world.

In California, the state with the most plant diversity in the U.S., and where over 200 endemic plants are known from fewer than five populations, eight gardens are members of CPC. Known as California's native garden, Rancho Santa Ana Botanic Garden is the largest botanical garden dedicated to California's native plants. One Endangered species in the care of Rancho Santa Ana is the small tree Catalina mountain mahogany (*Cercocarpus traskiae*). Endemic to Catalina Island, a single wild population exists, consisting of seven individuals in a canyon covering an area of approximately 2,700 square feet (250 m^2). Rancho Santa Ana holds vegetatively propagated representatives of all seven individuals, as well as an eighth cultivated plant which has been determined to be genetically different from the wild plants. Maintaining the complete set of known individuals provides the best hope for allowing the survival of this species in the wild by keeping future management options open.

The National Tropical Botanical Garden (NTBG) is one of several organizations concentrating on the conservation of Hawaii's rich flora. The Garden implements an integrated strategy that tackles threats and recovery needs at the species and landscape levels. It manages three botanic gardens in Hawaii together with areas of natural vegetation. Conservation strategies include plant collection and propagation in native plant nurseries, combined with ecological restoration and reserve management. Ex situ and in situ conservation techniques are integrated to help ensure the long term survival of the rare and threatened plants. Over the past 20 years, approximately 25 species have been rediscovered by NTBG botanists that were thought to be extinct.

Concentrated conservation efforts are a top priority for around 230 Hawaiian plant species that have fewer than 50 individuals remaining in the wild. The Hawaii Plant Extinction Prevention (PEP) Program is working diligently to preserve these extremely rare plants through controlling threats and monitoring plants in situ, maintaining ex situ collections, and searching for new individuals. NTBG, a PEP partner, has developed strict protocols for collecting very rare plants from the wild as part of their recovery work. The protocols include genetic sampling of populations, recording accurate locations for individual plants and populations, and the creation of high-quality species distribution and survey maps. When the plants are brought into cultivation, they are propagated by whatever means possible, to ensure that the number of individuals can be increased. If seed is available, this is stored to ensure the long term ex situ conservation of the severely threatened species.

CPC provides guidelines to ensure that plant conservation is carried out to very high standards. This is extremely important when working with plants that are close to extinction. Building collections of any species to directly support reintroduction efforts requires a significant investment in time, expertise, and resources. The potential conservation value of an ex situ collection depends on multiple factors. The type of plant material curated is of fundamental importance. In general, seed bank collections are best able to affordably secure genetically diverse material for direct conservation applications. However, seed banking is not yet possible for so-called recalcitrant species, such as oaks and other U.S. trees including walnut, yew (*Taxus*), and *Torreya* with seed that cannot be dried and frozen for long term storage, or species such as Stern's medlar that do not produce seed. Living plants or various forms of tissue culture may be the only viable option for high-value conservation collections. Genetic factors are also important in determining the success of ex situ collections based on the sampling protocols followed for acquiring plant material. For the best conservation results, well-documented, first-generation, wild-collected material that has been sampled to capture broad genetic diversity should be used. The maintenance and accessibility of viable and genetically diverse plant material are also crucial considerations. Without adequate curatorial and horticultural management, the conservation value, and the collection itself, can be entirely lost. Management protocols developed in zoos for maintaining genetic diversity are now being trialed for plants in botanic garden collections (Fant et al. 2016) to help avoid genetic erosion of the living collection. Sharing information and making plant collection material accessible to the botanical, conservation, and research communities as well as land managers is critical.

Another organization devoted to supporting the conservation work of botanic gardens is BGCI, a global organization of botanic gardens with over 90 members in the U.S. In addition, the American Public Gardens Association (APGA) is a broader organization representing the interests of public gardens in the U.S. APGA programs protect and recognize plant collections (Plant Collections Network), respond to the threats posed by invasive species (Plant Protection Program), and address the challenges of limited natural resources and climate change (Climate & Sustainability Alliance). The Plant Collections Network coordinates botanic garden germplasm preservation and promotes excellence in plant collections management. Collectively, CPC, APGA and BGCI work closely together and are mutually reinforcing in plant conservation. Working with federal land managing agencies through the Plant Conservation Alliance (PCA), as discussed in Collaborative Action below, ensures a coordinated approach to the complex issues of conserving plant diversity in the U.S. The North American Botanic Garden Strategy for Plant Conservation, 2016–2020 was published in 2016, and sets out how botanic gardens and their network organizations can work collectively to achieve a set of essential targets.

As mentioned above, a number of leading botanic gardens in the U.S. maintain important seed banks, storing native seeds as a long-term insurance policy and to support ecological restoration. In the U.S., the largest seed bank is maintained by the USDA Agricultural Research Service as part of The National Center for Genetic Resources Preservation (NCGRP) in Fort Collins, Colorado. The USFS National Seed Laboratory contributes seed to the NCGRP. The emphasis of the NCGRP has been on conserving the genetic diversity of crop species. However, since 2001, BLM and its partners have collected over 24,000 native wild plant species and have deposited them in the NCGRP seed bank. BLM works with federal agencies and non-federal partners on Seeds of Success, a national native seed collection program. This work underpins the development of native plant materials for restoration which is discussed in Chap. 6. The emphasis is on "workhorse" plant species that are essential ecological components needed to sustain and restore native plant communities.

Box 4.5: Oak Storage Research at Fort Collins

Seedbanks are a cost effective and efficient way to store plant material for species conservation and ecological restoration. Unfortunately, however, not all plant species produce seed that can be stored by drying and maintaining at low temperatures, typically −1 °F(−18 °C). The seeds of such plants are described as recalcitrant. Acorns, the seeds of oak species, fall into this category. The water content of acorns is too high and the seeds are too large to withstand the rapid cooling necessary for conventional seedbanking. Cryopreservation techniques using surgically removed embryonic axes are, however, proving to be successful for germplasm conservation for rare oak species at NCGRP in Fort Collins, CO.

Source: Walters et al. (2016)

Collaborative Action

The Plant Conservation Alliance (PCA) is a consortium of organizations and agencies working to advance plant conservation in the United States. The PCA is represented by twelve federal government member agencies and over 380 non-federal cooperators. In 1995, the PCA developed a National Framework for Progress in Plant Conservation (National Framework) to provide a coordinated approach to plant conservation in the U.S. The National Framework consists of six broad strategies, and outlines 30 supporting goals and suggested actions to guide efforts for implementing a national plant conservation strategy at national, regional, and local levels. Since 1995, the PCA's federal members and non-federal cooperating organizations (including botanic gardens, universities, states, and others) have worked toward the 30 common goals outlined in the Framework to "*protect native plants by ensuring that native plant populations and their communities are maintained, enhanced, and restored*." PCA (2019)

The National Framework aligns with the targets of the North American Botanic Garden Strategy for Plant Conservation, 2016-2020, which aligns with the Global Strategy for Plant Conservation that is linked to the Aichi Targets of the UN's Strategic Plan for Biodiversity. The challenge for the PCA and its associated organizations is to raise the profile of plant conservation so that it is recognized as being of fundamental importance for security and well-being of humankind.

References

Botanic Gardens Conservation International (BGCI) (2019) Global distribution of botanic gardens mapping applet. https://www.bgci.org. Accessed 11 June 2019

CNPS (2017) CNPS De-Extinction Project. California Native Plant Society Blog. https://grownatives.cnps.org/2017/03/30/cnps-de-extinction-project. Accessed 11 June 2019

Fant JB, Havens K, Kramer AT, Walsh SK, Callicrate T, Lacy RC, Maunder M, Hird Meyer A, Smith PP (2016) What to do when we can't bank on seeds: what botanic gardens can learn from the zoo community about conserving plants in living collections. Am J Bot 103:1–3

Havens K, Kramer AT, Guerrant EO Jr (2014) Getting plant conservation right (or not): the case of the United States. Int J Plant Sci 175(1):3–10

Hoover H (1933) Proclamation 2032. March 1, 1933. Saguaro National Monument - Arizona by the President of the United States of America. https://quod.lib.umich.edu/p/ppotpus/4731703.PROC.001?rgn=main;view=fulltext. Accessed 11 June 2019

Kramer A, Havens K (2009) Plant conservation genetics in a changing world. Trends Plant Sci 14:599–607

Kramer A, Hird A, Shaw K, Dosmann M, Mims R (2011) Conserving North America's threatened plants. Progress report on Target 8 of the global strategy for plant conservation. Botanic Gardens Conservation International US, Glencoe, IL, p 48

Kutner LS, Morse LE (1996) Reintroduction in a changing climate. In: Falk DA, Millar CI, Olwell M (eds) Restoring diversity. Strategies for reintroduction of endangered plants. Island Press, Washington, DC

Liebesman L, Lake E, Landreth P (2009) The endangered species act and climate change. Environ Law Rep 12:11173 11180

NatureServe (2019) NatureServe actionable knowledge to sustain biodiversity. https://www.natureserve.org. Accessed 11 June 2019

Plant Conservation Alliance (PCA) (2019) Plant conservation alliance. https://www.blm.gov/pca. Accessed 11 June 2019

Roosevelt T (1909) Special message of the President transmitting the Report of the National Conservation Commission. In: Gannett H. Report of the National Conservation Commission. Volume I. National Conservation Commission, Washington, DC

State of Hawaii (2019) Division of Forestry and Wildlife, NativeEcosystems Protection and Management. Rare plant program https://dlnr.hawaii.gov/ecosystems/rare-plants/. Accessed 11 June 2019

Stein BA, Gravuer K (2008) Hidden in plain sight: the role of plants in state wildlife action plans. NatureServe, Arlington, VA

Still SM, Frances AL, Treher A, Oliver L (2015) Using two climate change vulnerability assessment methods to prioritize and manage rare plants: a case study. Nat Areas J 35(1):106–121

Still SM, Havens K, Vitt P (2016) Assessing the vulnerability to climate change for rare plants in the western United States. Report to Bureau of Land Management Plant Conservation Program. Washington, DC

USGS (2019) Gap analysis project. https://www.usgs.gov/core-science-systems/science-analytics-and-synthesis/gap. Accessed 11 June 2019

US Geological Survey (USGS) (2018) Protected areas database of the United States (PAD-US) 2.0. https://www.sciencebase.gov/catalog/item/5b030c7ae4b0da30c1c1d6de. Accessed 11 June 2019

Walters C, Hill L, Crane J, Michalax M, Ke X, Carstens J, Conrad K, Westwood M, Colwell A, Clines J, Chmielarz P (2016) Preserving oak (*Quercus* sp.) germplasm to promote ex-situ conservation. Int Oaks J Int Oak Soc 27:255–266

Chapter 5
Land Ownership, Land Management, and Land Use

There was nothing but land: not a country at all, but the material out of which countries are made. (Willa Cather 1918)

The U.S. remains a land of plenty with abundant natural resources. Much of the land has been significantly altered, with over half highly modified or converted to agriculture or urban land use. The pace of development over the past 150 years has been remarkable. Development continues apace with about 44 million acres (18 million ha) of land newly developed between 1982 and 2010 (USDA 2015). How the land is allocated and managed has profound implications for economic and ecological well-being. Maintaining and managing land for biodiversity conservation supports ecosystem services needed for agriculture and other forms of land use. Implementation of environmental legislation and policies agreed upon for the common good depends to a significant extent on who controls the land. This chapter discusses current and historical land ownership, management, and land use within the U.S. Whereas much of the land remains fertile and productive, there are increasing areas of degraded land. Information presented here provides a backdrop to the ecological restoration activities that are increasingly taking place to repair the green infrastructure. Restoration is needed to repair the damage caused by lack of ecological understanding in the past and to address the current increasing pace of environmental change. Ecological restoration is discussed more fully in Chap. 6.

Land Ownership

Land, traditionally managed by Native Americans, provided the goods and services needed to sustain their people for many centuries. Land was as endless as the sky and was common property with no concept of private ownership until the arrival of European colonists. The first deed of Indian land to English settlers took place in Massachusetts in 1625. The extension of land ownership continued, based on

S. F. Oldfield et al., *Seeds of Restoration Success*, Springer Earth System Sciences, https://doi.org/10.1007/978-3-319-96974-9_5

English law precedents with its strong private property rights as advanced by the philosopher John Locke (1632–1704). Private property rights were seen as preventing the concentration of power in the hands of government and encouraging wealth generation. Early colonial charter companies allocated private ownership of land to each founding family. These parcels of land were eventually subdivided by inheritance and transfer, creating lots for different purposes, whether agricultural, commercial, or residential (Nolon 2006).

At American independence, the thirteen royal, charter and proprietary colonies established by the British became the first states of the U.S., extending from Massachusetts to Georgia. Subsequently, the Ordinance of 1784 created a framework for governance of land west of the Appalachian mountains and north of the Ohio River and also explicitly guaranteed property rights. The Northwest Ordinance of 1787 related to land of the entire Midwest region located east of the Mississippi River, laying down provisions for the establishment and governance of new states. This Ordinance guaranteed that newly created states would be equal to the original thirteen states. Following English law, it provided for land transfers between generations both by will and by contract, with provisions that took into account the frontier difficulties in registering deeds with distant officials. Over time, the combination of European ideas of natural rights, the adapted English common law of property, and American conditions of plentiful fertile land led to the distribution of land into private ownership with secure titles, and this formed the basis for the expansion of a free society westward (Morriss 2007).

Federal ownership of land began when the original thirteen states of the new nation ceded title to over 40 percent of their territory between the Appalachian Mountains and the Missouri River to the central government between 1781 and 1802. Federal land was subsequently acquired by major transactions with foreign governments. The Louisiana Purchase of 1803 acquired land stretching from the Mississippi to the Rocky Mountains from the government of France. Land was acquired from Britain when the boundary with Canada was agreed upon in 1818 and Florida was purchased from the government of Spain in 1819. Western land was added to the U.S. from Britain, through war with Mexico, and through purchase of Alaska from Russia in 1867. Public domain lands are primarily those obtained from a foreign country. They are governed by different laws than those covering land acquired from states or individuals (Alexander and Gorte 2007). Much of the land brought into federal ownership was subsequently sold to raise funds for the treasury or was allocated to former soldiers, but vast tracts of public land remain.

Currently almost 30 percent of U.S. land (640 million acres/260 million ha) is under public ownership with management by federal agencies. Most of these lands are in the West and in Alaska. Specifically, 61.2 percent of Alaska is federally managed, as is 46.9 percent of the 11 coterminous western states. By contrast, the federal government manages 4.0 percent of lands in the other 38 states. This western concentration has contributed to a higher degree of controversy over land ownership and use in that part of the country (Vincent et al. 2014).

Federal Land Management

Four federal agencies administer 608.9 million acres (246 million ha) of public land: the Bureau of Land Management (BLM 2019), U.S. Fish and Wildlife Service (FWS), and National Park Service (NPS), all within the Department of the Interior (DOI), and the Forest Service (USFS) in the Department of Agriculture. In addition, the Department of Defense administers 14.4 million acres (5.8 million ha). Various other agencies administer the remaining federal acreage. The four major land managing agencies manage these lands for a range of purposes, primarily related to recreation, conservation, and development of natural resources. Federal policies and laws that direct responses to issues such as the control of wildfire and the conservation of threatened species apply to all federal lands as shown in Fig. 5.1.

The BLM manages 245 million acres (100 million ha) of public land, more than any other federal agency, with land located primarily in the western states. It also has responsibility for about 700 million acres (283 million ha) of federal subsurface mineral estate throughout the nation, underlying both federal and non-federal lands (BLM 2019). The BLM was established in 1946 by combining the General Land Office, created in 1812 to distribute federal land, and the Grazing Service, responsible for implementing grazing policy on public land (Muhn and Stuart 1988). The Bureau has a multiple-use management mission, set out in the Federal Land Policy and Management Act of 1976. This mandates the management of public land resources for a variety of uses, such as energy development, livestock grazing, recreation, and timber harvesting. At the same time, protection of natural, cultural, and historical resources, is an important part of BLM's mission. Important conservation sites are found in the BLM's 34 million-acre (13.8 million ha) National Conservation Lands. This includes 224 Wilderness Areas covering a total of 8.7 million acres (3.5 million ha), as well as 16 National Monuments comprising 4.8 million acres (1.9 million ha). With the increasing occurrence and severity of wildfires, management of conservation areas and other landscapes, the rehabilitation of degraded ecosystems is becoming increasingly important on BLM lands.

The FWS was developed from the merging of a commission set up to manage fish stocks in 1871 with divisions of the Department of Agriculture that were established to survey and manage birds and mammals. The FWS now manages 89.1 million acres (36.1 million ha) of land, primarily to conserve and protect animals and plants. The National Wildlife Refuge System is a Division within the FWS with responsibility for land managed as wildlife refuges, waterfowl production areas, and wildlife coordination units.

The NPS was established in 1916 with a remit to protect the 35 National Parks and monuments managed by the Department of the Interior at the time and those declared in the future. It now manages 79.6 million acres (32.2 million ha) of land in 401 diverse units to conserve scenic lands and natural resources, together with historic sites, making these available for public use. Activities that harvest or remove resources generally are prohibited on NPS lands.

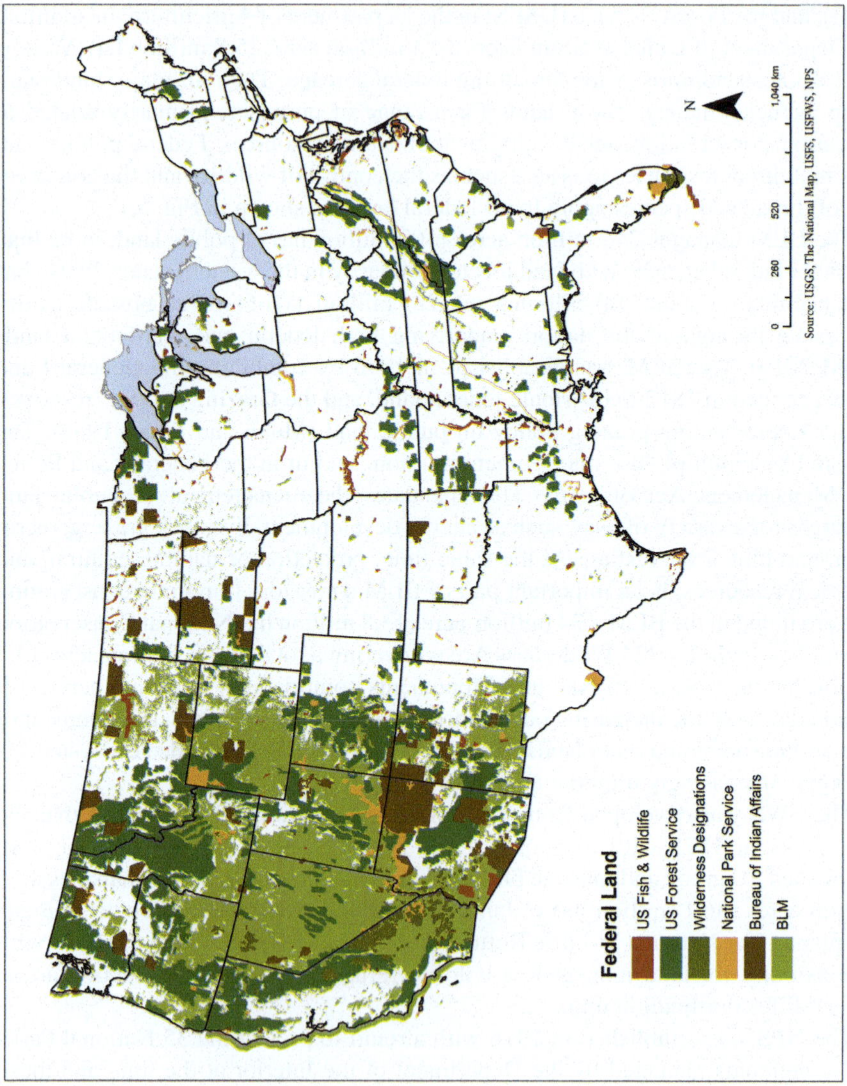

Fig. 5.1 Map of federal land management boundaries

The USFS was established in 1905 with responsibility to manage areas of land already designated as forest reserves. These reserves were envisioned as working forests with multiple objectives. The Organic Administration Act of 1897, under which most national forests were established, states:

> No national forest shall be established, except to improve and protect the forest within the boundaries, or for the purpose of securing favorable conditions of water flows, and to furnish a continuous supply of timber for the use and necessities of citizens of the United States… (USFS 2019).

Currently, the USFS manages 193 million acres (78 million ha) of land. Like the BLM, the USFS is required to manage land for multiple uses. Management for sustained yields of various products and services, including timber harvesting is one objective. Others are management for recreation, grazing, watershed protection, and maintenance of fish and wildlife habitats. Most of the USFS lands are designated national forests. Of these, approximately 73 percent of the 193 million acres (78 ha) of national forests are considered to be forested with 35 percent of the land available for regularly scheduled timber harvest. The remaining 65 percent of the forested land is designated for non-timber uses, such as wilderness preservation and recreation. Some of the forested lands cannot be harvested for timber because of environmental constraints such as rugged terrain and fragile soils. National Grasslands (see below) fall within the estate of the USFS. Wildfire protection is a major management responsibility and is becoming increasingly important for the USFS as well as the BLM (USFS 2019).

Box 5.1: Management of Land for Carbon Storage

Land management priorities have changed over time with changing patterns of food, timber and energy consumption. A relatively new requirement is the management of federal land to address climate change. Management and restoration of ecosystems to increase ecological carbon sequestration and reduce overall levels of greenhouse gas emissions are mandated in the Energy Independence and Security Act of 2007 (see Appendix 1). In developing restoration, management, and adaptation strategies as required under the Act, the use of native plant species (including mixtures of many native plant species) for sequestering covered greenhouse gases in each ecosystem is emphasized. The National Fish, Wildlife and Plants Climate Adaptation Strategy (see Appendix 1) mandated by Congress in 2009, recognizes that, "*Management plans in this new era of climate change must take into account the differing capacities of individual species and ecosystems to adapt to changing conditions and sustain cultural, recreational, and commercial uses.*"

As noted in Chap. 1, more than 30 federal agencies have developed Climate Change Adaptation Plans. Building on these plans, the Executive Order 13653 "Preparing the United States for the Impacts of Climate Change," (see Appendix 1) was signed in 2013. With regard to managing the environment, this Executive Order directs federal agencies to manage lands and waters for climate preparedness and resilience. This Executive Order was revoked in 2017.

(continued)

Box 5.1 (continued)

Management of land at a time of global climate change involves addressing uncertainty and the impact of extreme weather events. In the West, increased drought and incidence of wildfires has been a consequence of climate change. Responding to this in January 2015, the Department of Interior Secretarial Order on Rangeland Fire Prevention, Management and Restoration (SO3336) (see Appendix 1) called for enhanced action to prevent and suppress rangeland fire and to restore one particular ecosystem - fire-damaged sagebrush communities that occur extensively throughout the West.

Management of Non-federal Land

Non-federal lands in the U.S. include privately owned lands, tribal and trust lands, and lands controlled by state and local governments. According to the U.S. Census Bureau, there are about 39,000 county, municipal, and township local governments that have or can be given authority to regulate private land use (Nolon 2006). Land use zoning is a familiar form of regulatory power. The first zoning ordinance in the U.S. was passed in New York City in 1916, and, by the 1930s, most states had adopted zoning laws.

Non-federal rural lands are predominantly forest land (413 million acres/167 million ha), rangeland (406 million acres/164 million ha), and cropland (363 million acres/147 million ha) as defined in Box 5.2 (USDA 2015). There are frequent shifts in these land uses with changes impacting contiguity of habitat, hydrologic features of the landscape, erosion potential, and other natural processes or functions (USDA 2015).

Non-federal land management is influenced by government policy in a variety of ways. Agricultural land management is, for example, influenced by farming commodity price-support programs and federal crop insurance. Federal spending programs provide incentives to state and local governments and private landowners to achieve environmental objectives of national importance (Nolon 2006). With regard to conservation, the Conservation Reserve Program (CRP) is particularly relevant. This federal program was established under the Food Security Act of 1985 and was developed initially to address soil erosion, a significant national problem in the 1970s. The CRP assists private landowners in converting cropland and pasture that is environmentally sensitive and liable to erosion to vegetative cover for 10 years or more. Incentives are available to landholders in areas of agricultural production, buffer areas, and wetlands across the country, inducing them to enter into contracts to change land use. Eligible lands must be liable to high levels of erosion or in a state or national conservation priority area. Landowners and operators with eligible lands compete nationally for acceptance during specified enrollment periods. Another form of CRP was introduced in the 1996 Farm Bill. Eligible lands must be suitable to serve as one of a number of high priority conservation practices, such as a wetland restoration, filter strip, riparian buffer, or field windbreak. Enrollment of eligible lands is continuous and without competition.

Since CRP began in 1986 it has achieved major environmental gains. The program has reduced more than 8 billion tons (7.3 billion metric tons) of soil erosion, achieved the annual equivalent of approximately 57 million tons (52 million metric tons) net reduction in carbon dioxide from sequestration, reduced fuel use, reduced the application of nitrogen and phosphorus and reduced nitrous oxide emissions from fertilizer use. At the same time more than 2 million acres (0.8 million ha) of wildlife habitat has been established (Stubbs 2014).

Box 5.2: Land Cover/Use Categories in the U.S.

Rangeland

Land that still contains elements of natural vegetation on which the climax or potential plant cover is composed mainly of native grasses, grass-like plants, forbs or shrubs suitable for grazing and browsing. The vegetation also includes introduced forage species that are managed like rangeland. This includes areas where introduced hardy and persistent grasses, such as crested wheatgrass (*Agropyron cristatum* complex), and various management practices are used with little or no chemicals or fertilizer being applied. Grasslands, savannas, many wetlands, some deserts, and tundra are considered to be rangeland. Certain communities of low forbs and shrubs, such as mesquite, chaparral, mountain shrub, and pinyon-juniper, are also included in this category.

Forest land

Land over 1 acre (0.4 ha) in size that is at least 10 percent stocked by single-stemmed woody species that will be at least 13 feet (4 m) tall at maturity. Forest land also includes land showing evidence of natural regeneration of tree cover (cut over forest or abandoned farmland) and not currently developed for non-forest use. Ten percent stocked equates to an areal canopy cover of leaves and branches of 25 percent or greater.

Cropland

Land used for the production of adapted crops for harvest. Two subcategories of cropland are recognized: cultivated and noncultivated. Cultivated land comprises land in row crops or close-grown crops and also other cultivated cropland; for example, hayland or pastureland that is in a rotation with row or close-grown crops. Noncultivated cropland includes permanent hayland and horticultural cropland.

Pastureland

Land managed primarily for the production of introduced forage plants for livestock grazing. Pastureland cover may consist of a single species in a pure stand, a grass mixture, or a grass-legume mixture. Management usually consists of cultural treatments: fertilization, weed control, reseeding, renovation, and control of grazing.

Source: USDA (2015)

Land trusts provide an important mechanism for conservation of private lands, either by ownership or by negotiating conservation easements. Ownership of reserves by land trusts requires management of natural resources and monitoring of progress toward stated management objectives (Peters et al. 2017). A conservation easement is a legally-binding agreement between a property owner and a nonprofit organization, usually a land trust, or a government agency that restricts development on an area of land, usually in exchange for tax benefits for the property owner (Gattuso 2008). Various types of private land use, such as farming, ranching and timber harvesting, can continue under the terms of a conservation easement, and owners may continue to live on the property. The organization to receive or buy the easement, known as the grantee, holds interest in the property and enforces the restrictions. Most conservation easements are binding in perpetuity.

Land trusts in the U.S. currently manage over 56 million acres (22.7 million ha) of land (Chang 2016). The land trust movement began in 1891 in the State of Massachusetts with the first private land trust, the Trustees of Reservations in Massachusetts. This was formed to purchase and maintain public parkways, designed by Frederick Law Olmsted, throughout the city of Boston. Conservation easements were next used in the 1930s and 1940s, when the NPS bought parcels of land for scenic use along what are now the Blue Ridge Parkway and the Natchez Trace Parkway. Massachusetts, with 143 land trusts, is still the nation's leading state for land trust management.

Four national land trusts are The Nature Conservancy Trust, Trust for Public Land, American Farmland Trust, and Rails-to-Trails Conservancy. Most of the Nature Conservancy's easement transactions have been donated with the Conservancy also purchasing some easements in high priority sites. Purchasing easements is less expensive than purchasing the land itself. The Nature Conservancy also supports purchase-of-development rights (PDRs) programs. These are a form of conservation easement whereby public funds are used to purchase the land's development rights, in effect creating a conservation easement on the land while the land remains in private ownership. Increasingly, states and communities are using financing mechanisms for PDR programs such as annual appropriations, dedicated lottery revenues, and bonds authorized by the legislature or a voter referendum.

Management of Land for Agriculture

Currently around half the land in the U.S. is managed for agriculture, including cropland, pasture, range and grazed forestland. Over 60 percent of land in farms is owner-operated, a proportion that has been relatively stable over the past 50 years. The remaining land is rented. More than half of cropland is rented, compared with just over 25 percent of pastureland. In general, rental activity is concentrated in areas dominated by production of crops such as rice, corn, soybeans, wheat, and cotton. The form of land tenure impacts land management decisions with landlords more likely to be involved in long-term decisions, such as adopting permanent conservation practices and participating in Government programs (Bigelow et al. 2016).

The development of modern agriculture in the U.S. has resulted in fundamental ecological changes as elsewhere in the world, with clearance of forests, modification of prairies, drainage of wetlands and irrigation of arid areas. At the same time many new species have been introduced into the landscape either intentionally or accidentally.

Prior to European settlement in eastern areas of the U.S., Native Americans cleared and farmed modest areas of land mainly along river floodplains and seasonal settlements. Early European settlers arriving in New England found a wooded landscape with broad river valleys lined by sedge beds and floodplain grasslands. The marsh vegetation provided colonists with essential grazing for introduced cattle and sheep and provided hay and fodder for winter. Settlers progressively cleared uplands to provide productive farms, but the relatively few native grasses could not withstand grazing and trampling. Meadow plants, such as bluegrass (*Poa pratensis*) and white clover (*Trifolium repens*), were introduced from England to replace the native species for grazing. By the late seventeenth century "English grass" was one of the three major land types in Massachusetts and was of high value (Peterken 2013). Other species characteristic of England lowland haymeadows were introduced and persist to this day transforming elements of the U.S. native vegetation.

In the Pacific Northwest, later European settlement led to the introduction of European grassland plants on a huge scale. Grasses such as common velvetgrass (*Holcus lanatus*), annual velvetgrass (*Anthoxanthum aristatum*) (soft brome (*Bromus hordaceus)* and perennial ryegrass (*Lolium perenne*) became established in response to the new practices of fire-suppression and year-round heavy grazing by cattle and sheep. Cheatgrass (*Bromus tectorum*) is a species native to the Mediterranean region. Part of its original habitat was the decaying straw of thatched roofs. Introduced into the U.S. in packing materials, ship ballast, and likely as a contaminant of crop seed, cheatgrass was first found in the U.S. near Denver, Colorado, in the late 1800s (Whitson 1991). By the 1930s, cheatgrass was becoming the dominant grass over vast areas of the Pacific Northwest and Intermountain West. Cheatgrass developed into a severe weed in several agricultural systems throughout North America, particularly western pastureland, rangeland, and winter wheat fields. It is now estimated to infest more than 101 million acres (41 million ha) in western states (Mack 1981).

Originally concentrated in the coastal plains, the spread of modern farming across the U.S. occurred rapidly during the nineteenth century. Westward migration became significant when Daniel Boone was commissioned by the Virginia Legislature to build the "Wilderness Road" through the Cumberland Gap in the Appalachian Mountains. From 1775 to 1810 around 300,000 settlers traveled across the mountainous route heading for the low hills and river bottom lands of central Kentucky and beyond. When steam boat traffic started on the Ohio River, travel largely left the overland route, and the mountain section was nearly isolated from the outside world.

After 1870, new railroads offered attractive packages of land and transportation to American farmers and European immigrants, who also took advantage of the Homestead Act of 1862. This law granted 160 acres (65 ha) of government land, free of charge on the condition that it was farmed for at least 5 years. Nearly 1.5 million people acquired over 309,000 square miles (800,000 km^2) of land, primarily in the Great Plains region. This Act, together with various other land transactions, resulted in a major conversion of prairie to agriculture. The most significant impact was in the tallgrass prairie where the deep soils were well suited to arable crop production. Much of the Great Plains became rangeland with free-ranging cattle controlled through spring and fall roundups. Such ranching began in Texas and gradually moved northward with cattle herded north to rail heads such as Dodge City, Kansas and Ogallala, Nebraska.

The open-ranching economy collapsed rapidly in the late 1880s. Widespread overgrazing, competition from the superior beef of expanding cattle-raising operations in the Midwest, a falling national economy, a disastrous winter in 1887–1888, and a rapid influx of farmers onto the Plains combined to curtail this form of land use. The open-range, unimproved ranches were pushed to the drier western side of the Plains or were forced into a more restrained fenced operation. In 1895, Jared G. Smith of the U.S. Botanical Survey, noted the extreme depletion of range in the Great Plains. He recognized the erosion and compaction of soil, increase in poisonous and woody plants as symptoms of range deterioration (Holechek 1981).

On the agricultural frontier, barbed wire, developed commercially in the 1870s, provided an effective alternative fencing material, replacing wood that was in short supply in the Prairies. The introduction of windmill and mechanical well-drilling devices enabled better management of local water supplies. Grain farming also became increasingly mechanized, enabling farmers to operate larger farms and thus compensate for lower yields. Crops such as hard winter wheat (*Triticum aestivum*) that were better adapted to the growing conditions of the region were introduced. Hard winter wheat was first brought to the U.S. by Mennonite immigrants from Russia. The agriculture of the Great Plains today remains large scale and mechanized, dominated by a few crops, the most important of which is still wheat.

The acquisition of Native American lands by the federal government during the nineteenth century, and its distribution to settlers, led to the creation of reservations. These reservations were often in areas ill-suited to agriculture, particularly to the agricultural techniques introduced by missionaries and government employees. Wheat and row cultivation were promoted over the wider range of traditional crops and intercultivation methods that had traditionally been used. By the 1880s, pressure on reservation lands from settlers became acute. The General Allotment Act (also known as the Dawes Act) in 1887 provided for the survey of tribal land and allocation of land to individual tribal members. Each head of household received a plot of land in accordance with the Act, generally 160 acres (65 ha), leaving the remainder of the reservations to be sold as surplus lands. Individuals who took allotments would receive title to their land after a trust period of 25 years (Wishart 2004).

U.S. agriculture flourished in the early years of the twentieth century but declined during the 1920s. The stock market crash of October 1929 led to falling incomes for farmers, and its impacts in following years coincided with the prolonged drought that covered the Corn Belt and southern Great Plains. Lasting nearly a decade, and considered to be the worst environmental disaster in the U.S., many crops were damaged by lack of rainfall, high temperatures, and high winds, as well as insect infestations and dust storms. At its worst, the Dust Bowl covered about 100 million acres (40 million ha) in the Southern Plains. "*The grasslands that evolved to hold the soil in place and sustain regional species were destroyed within just a few decades by the expansion of cash crop agriculture and ranching following the opening of the Plains to white settlement*" (Holleman 2017). Poor land management practices contributed to the economic and social consequences of the drought, which sent shockwaves throughout the country. Millions of people migrated from the drought areas in search of new livelihoods. As they headed west, conflicts arose with longer-established residents, with relief and health agencies overstretched.

The 1936 Report of the Great Plains Drought Committee asserted that the basic cause of the Dust Bowl disaster was the imposition of a system of agriculture suited for humid regions on an area with semi-arid conditions. Misguided land-allotment practices under the 1862 Homestead Act were implicated. Another factor was lack of knowledge and experience on the part of settlers who were misled by government policies encouraging "*a system of agriculture which could not be both permanent and prosperous*" (Great Plains Drought Committee Report 1936). The problem was intensified by mechanization and the highly speculative nature of wheat farming resulting from the combined forces of nature and the market, including fluctuations in prices during and after World War I (Holleman 2017).

It is impossible to count all the costs associated with the 1930s drought, but one estimate by Warrick (1980) claims that financial assistance from the government may have been as high as $1 billion by the end of the drought. As a consequence of the disaster, the government took steps to reduce the vulnerability of the regions to future droughts. The Soil Conservation Service (SCS), now the Natural Resources Conservation Service, emphasized soil conservation measures such as terracing and contouring. The first soil conservation districts came into being. National Grasslands were created by legislation in 1937, which authorized the federal government to acquire damaged agricultural lands for rehabilitation. Of the twenty National Grasslands, all are on or at the edge of the Great Plains with the exception of three in southeastern Idaho, northeastern California, and central Oregon. The total area of all 20 National Grasslands is 3.8 million acres (1.6 million ha).

Despite efforts to restore grasslands and conserve soil through improved farming methods, the land of the Great Plains is still not completely restored. Once thriving farms are still abandoned, and new dangers such as excessive extraction of water for cotton cultivation together with the changing climate are again putting the region at serious risk (University of Nebraska National Drought Mitigation Center 2019).

It is generally recognized that over the next 40 years agricultural practices will need to become increasingly flexible globally to provide for food security and respond to environmental change. The multifunctional nature of crop production and livestock management will need to deal with change and uncertainty. Resilience and adaptability will become more important considerations. To achieve this, production systems will need to have greater reliance on ecological processes that produce positive feedbacks on sustainability and production and ensure improved provision of all ecosystem services (FAO 2019). These changes will involve integrated use of agricultural biodiversity, bringing together the contributions of crops, livestock, agroforestry species, soil organisms, pollinators, and other components (FAO 2011).

Box 5.3: Land Degradation in California

As noted in Chap. 2, California is a global biodiversity hotspot with diverse ecosystems and an abundance of endemic species. Over 30 percent of the native plants are endemic to the state. Human impacts have resulted in California being one of the four most ecologically degraded states in the U.S. The gold rush, construction of railroads, development of major cities and rapid expansion of cattle and sheep production in the second half of the nineteenth century all contributed to rapid changes in the landscape. Many of California's rare plants have become endangered as a direct result of development over the past 150 years. All or part of the eight most threatened U.S. ecosystems are represented in California: beach and coastal strand, southern California coastal sage scrub, large streams and rivers, California riparian forests and wetlands, California native grasslands, old-growth ponderosa pine (*Pinus ponderosa*) forests, cave and karst systems, and the ancient forests of the Pacific Northwest.

Native grasslands and vernal pool (seasonal wetland) habitats in California have been reduced to about 1 percent of their original extent by the conversion of natural lands to agricultural fields and livestock pasture, urban development, and the invasion of exotic grasses. Introduced plant species comprise over 90 percent of plant cover in most California grasslands. Nevertheless these grasslands provide habitat for nearly 90 percent of state-listed rare and endangered species and 75 federally listed plants and animals. They also provide 75 percent of the state's livestock forage. These grasslands are critical in regulating water flow and water quality, and contribute significantly to regional soil carbon storage. They also maintain populations of pollinators needed to support California's agriculture. Conservation and restoration of the state's grasslands largely depends on private landowners as 88 percent are privately owned.

Rangeland in general, including natural and seminatural grasslands, savannas, and shrublands managed for grazing, covers about a third of the state. After more than 200 years, grazing remains California's most extensive land use, with goals and practices strongly affected by patterns of land ownership. During recent decades, sheep and beef cattle production have slowly declined. California continues to lose its rangelands to urbanization that fragments habitat and disrupts the water cycle. Protecting rangeland and ranching has become a shared priority for different interest groups including ranchers, conservationists, open space organizations, and municipal utility districts.

(continued)

Box 5.3 (continued)

Other seriously threatened ecosystems include wetlands, riparian woodlands, and southern maritime sage scrub, which have all been reduced to 10 percent or less of their original area. Wetlands are destroyed by land filling and the diversion of water for agricultural, industrial, and residential development. Riparian forests face threats from logging, grazing, and development, while coastal sage scrublands are threatened by housing and commercial development, and the growing use of off-road vehicles. Redwood forests have been reduced by intensive logging to 15 percent of their original area during the last 150 years.

More than 30 percent of California's plant species are threatened with extinction. Fortunately, however, considering the scale and magnitude of agricultural development, suburban sprawl, water diversion, resource extraction, pollution, and introduction of over 1,500 invasive species, relatively few species have become extinct. The state's systems of protected areas cover 46 percent of the land area, helping to ensure the survival of species in their natural habitats.

Sources: Larson-Praplan (2018), Mooney H & Zavaleta (2016)

Management of Land for Forestry

The amount of forestland in the U.S. has been relatively stable for the past century. Prior to European settlement, forests covered about 1 billion acres (0.4 billion hectares), nearly half the total land area. The vast areas of forests and their rich underlying soils were of critical importance in the development of the U.S. From 1600 to 1900, about 25–30 percent of the original forest area was converted to other land uses, primarily agriculture. Nearly two-thirds of the net conversion to other uses occurred in the second half of the nineteenth century, when an average of 13 square miles (34 square km) of forest was cleared every day for 50 years. By 1910, forest land had declined to an estimated 754 million acres (305 million ha), or 34 percent of the total land area. The amount of forestland in the U.S. now comprises 766 million acres (310 million ha), or 33 percent of the total land area. (Oswalt and Smith 2014).

The way forestland is defined affects the total forest acreage reported. Recently, the USFS adopted new standards for what qualifies as forestland. For example, ecosystems such as chaparral and sparse pinyon-juniper "forests" (with 5-10 percent tree cover) in the Rocky Mountains and West Coast have been reclassified as woodlands rather than forestlands. The USFS estimates that the new definition removes about 6–8 million acres (2.4-3.2 million ha) previously considered forestlands. Generally, these changes do not affect historic estimates of productive, unreserved forestland or timberland.

Increases in forest in the twentieth century have resulted from natural regeneration of marginal or abandoned agricultural land, for example in New England, and from increasing reforestation efforts. Although forest cover has increased, the composition and structure has changed significantly. In the Northeastern states for example, 80 percent of the region was forested in 2010, but with less than 1 percent of old-growth forest remaining intact (Thompson et al. 2013). The composition of forests in the region has changed with a broad ecological shift away from species such as beech (*Fagus*) and hemlock (*Tsuga*), to the now dominant maple (*Acer*) and poplar (*Populus*) species.

Logging by European settlers began in the rich forests of the northeast, providing an immense timber resource for export and local use. Logging expanded into the Appalachians with improved transport links. Within the Appalachian Mountains, large areas of forest land with timber stocks remain today despite scattered logging operations that were active near the river courses as early as 1825. Following the Civil War, deposits of coal and iron were discovered in the Appalachians and railroads opened up the area for the timber industry. Loggers from New England, New York and Pennsylvania took advantage of the new sources of timber, having moved through the virgin forests of the northeast. Timber mills were built throughout the mountains to process the huge hardwood trees for a wide variety of products.

Until World War II, the USFS primarily focused on watershed protection, forest restoration, and wildfire prevention and suppression. Because there were abundant supplies of private timber, very little national forest logging occurred during this period. During the post-World War II housing boom, national forests were viewed as a ready supply of building material. The increased demand for timber from national forests and congressional inducements led to more widespread use of intensive harvesting techniques such clearcutting. Along with the increased logging that followed, concern over the environment increased. In the 1960's and 1970's, laws were enacted to protect forests. Legislation formalized the concept of "multiple-use," whereby the uses of timber, forage, and water shared equal footing with wildlife conservation and recreation opportunities. The National Forest Management Act (NFMA) of 1976 mandated public input into all forest management plans.

In the 1980s when initial attempts to stop logging of old growth and other forested areas proved unsuccessful, a series of lawsuits invoking protection of the habitat of the federally threatened northern spotted owl (*Strix occidentalis caurina*) ultimately led to logging cessation in major areas of federal lands in the western states (Manheim 2010). The 1994 Northwest Forest Plan (NWFP) shifted federal land management from an emphasis on timber production to ecosystem management and biodiversity conservation. The plan established a network of conservation reserves and an ecosystem management strategy for an area of over 38,000 square miles (98,000 square km) from northern California to Washington State within the range of the northern spotted owl. Subsequent assessments and monitoring programs established under the plan have demonstrated the effectiveness of this reserve network and ecosystem management approach (DellaSala et al. 2015).

As a result of concerns about environmental impacts and conflicting uses, increased lawsuits, and additional environmental protection measures, the USFS now operates federal timber sales under more effective environmental protection policies. In response to the public controversy and a greater understanding of how management actions influence the landscape, today's timber sale levels have dropped by two-thirds (back to the pre-1950 levels), even though timber demand continues to increase at a rate of about 1 percent annually. In addition, clearcut harvests have been reduced by 80 percent over the last decade.

Along with national forest programs, the USFS conducts considerable research aimed at improving the management of forests in an ecologically sound manner. In 1910, the U.S. Forest Service's Forest Products Laboratory was created with a specific mission of improving forest resource conservation. The laboratory has helped to substantially reduce wood use and industrial pollution through the development of wood composites (fiberboard, etc.), improved pulping operations, innovations in the use of recycled wood products, and many more.

Fifty-seven percent of U.S. forestland is privately owned by individuals or companies; 43 percent is public forestland managed by federal and state agencies. Over 60 percent of public forestlands are in the Rocky Mountain region and Alaska.

The area of U.S. forestland under conservation protection is almost double the world average. In U.S. private forests, conservation efforts are increasing nationwide through conservation easements, and 25 percent of private forests have been certified as being sustainably managed.

The legal framework for ensuring forest sustainability and conservation of forests is both comprehensive and complicated. Despite the many laws and regulations that govern forestry in the U.S., there is no overarching forestry policy that ensures sustainability. The majority of land use regulations are often not specifically forestry related but are concerned with more general environmental concerns, such as water or air quality. Major federal conservation statutes that help conserve and protect forest resources are included in Appendix 1.

Management of Land for Recreation and Conservation

Early concerns for the loss of natural resources focused on public forests. The settlers moving west from the Appalachians gave little thought to forest conservation as they cleared the land for farming. But by the end of the nineteenth century the need to conserve timber resources became apparent. Exploitation of forests led to concerns about devastation of the public domain. The General Public Lands Reform Law passed in 1891 allowed the President to reserve public land with timber and undergrowth from settlement.

To a certain degree, the early conservation movement was aesthetic in nature, influenced by writers and artists such as Henry David Thoreau, Ralph Waldo Emerson, and George Catlin. These influential creatives articulated a desire to preserve areas of unique scenic beauty, leading to the formation of Yellowstone, Yosemite, and Crater Lake National Parks. The artist George Catlin became aware that the balance of nature was being destroyed because robes made of fur from bison (*Bison bison*) had become very fashionable in the cities of the East. Catlin noted that the craze for bison robes would make the animals extinct. Instead of killing the animals and using nearly every part of them for food, or to make clothing and even tools, Native Americans were being paid to kill bison for their fur alone. Bison were also killed to deprive Native Americans of their food source.

Appreciation and promotion of the magnificence of nature also stimulated tourism with its economic returns to cities such as Colorado Springs and Las Vegas. The national park movement was also an expression of the nationalistic spirit of the U.S. following the Civil War, hastened by rapid urbanization and industrial development. The growth of cities and changes in the landscape led to calls for the preservation of wilderness values. There were practical considerations too. Privately owned city utilities such as water supplies and city parks were brought under public ownership to ensure better service and sustainability.

Contributing to the growing appreciation and acceptance of the need for conservation was the increasing scientific awareness of natural history through geographical and botanical exploration and the landscape planning movement. The waste of natural resources was recognized by George Perkins Marsh, John Muir, and Frederick Law Olmsted. By the late nineteenth century, many Americans understood the complex relationship between soils, water and forests.

The first national park created in the U.S. was Yellowstone, established in 1872. Earlier, the Yosemite Valley in California was declared a state park under the Yosemite Bill signed by Abraham Lincoln in 1864, and it became a national park in 1890. On September 25, 1890, President Benjamin Harrison signed legislation establishing America's second national park. Created to protect the giant sequoia (*Sequoiadendron gigantum*) trees from logging, Sequoia National Park was the first national park formed to protect a living organism. One week later, General Grant National Park was created and Sequoia National Park was enlarged.

Several national forests were created under the Weeks Act of 1911 to restore forests on formerly private lands that had been heavily logged or cleared for agriculture. That law authorized the Secretary of Agriculture to "*…examine, locate, and purchase such forested, cutover, or denuded lands within the watersheds of navigable streams as in his judgment may be necessary to the regulation of the flow of navigable streams or for the production of timber.*" (USFS 2011) Many of today's Eastern national forests were acquired under the Weeks Act. Their healthy condition today can be attributed to past reforestation efforts by the U.S. Forest Service and partners such as the Civilian Conservation Corps.

Approximately 13 percent of the U.S. land is managed for conservation purposes under some form of national, regional, or international designation, with over 34,000 protected areas (UNEP WCMC 2018). Including conservation easements and similar forms of conservation agreements significantly increases the area of land conserved. The protected areas of the U.S. are a tremendous asset for biodiversity conservation. Their role in the *in situ* conservation of plant species is discussed in Chap. 4. A recent paper noted that the efforts for both private land protection and public protected areas are not prioritizing areas of high species endemism, or are having less success in those areas. Southeastern states with mostly private land have, for example, relatively few conservation easements (Jenkins et al. 2015).

Urbanization

Major movements of people in the U.S. have been divided into three periods. Initially, as described in this chapter, the settlement frontier spread from east to west as more and more farmlands became available. This was followed by a major surge in urban employment generated by the Industrial Revolution, with people moving from rural areas to towns and cities. From 1830 to 1930, the pace of urbanization in the U.S. accelerated substantially with the share of the population living in an urban area increasing from 10 percent to 60 percent. By 1990, this figure had reached 75 percent. The relative decline in rural population was matched with an absolute decline in the population of people living on farms. Between 1960 and 1987, for example, the farm population fell from more than 15 million to under 6 million. The third, and present broad period of U.S. population mobility is one of long-distance movement, predominately between metropolitan areas. California is the most highly urbanized state in the U.S. with almost 95 percent of its population living in urban areas.

The shift from rural to urban areas has profound implications for management of the land and biodiversity. Intensification of farming may free up more land for biodiversity conservation, recreation, and tourism but also concentrates environmental pressures in areas where it is practiced. Urban areas may themselves be highly important for biodiversity, but rapid urban development continues to be associated with air and water pollution and changes in fire frequencies. The shift to cities may result in people losing connections with the wild. Taking inspiration from nature to develop biodiverse planting schemes in urban settings is one of the themes we develop further in Chap. 7.

References

Alexander K, Gorte RW (2007) Federal land ownership: constitutional authority and the history of acquisition, disposal and retention. CRS Report for Congress RL34267, Washington, DC, 14 p

Bigelow D, Borchers A, Hubbs T (2016) US farmland ownership, tenure and transfer. EIB-161, U.S. Department of Agriculture, Economic Research Service, August 2016. 53 p

Bureau of Land Management (BLM) (2019). https://www.blm.gov Accessed 12 June 2019

Cather W (1918) My Antonia. Houghton Mifflin, Boston, MA

Chang K (2016) 2015 National Land Trust Census Report. Our common ground and collective impact. Land Trust Alliance, Washington, DC

DellaSala DA, Baker A, Heiken D, Frissell CA, Karr JA, Nelson SK, Noon BR, Olson D, Strittholt J (2015) Building on two decades of ecosystem management and biodiversity conservation under the Northwest Forest Plan. USA Forests 6(9):3326–3352

FAO (2011) Biodiversity for food and agriculture. Contributing to food security and sustainability in a changing world. Outcomes of an expert workshop held by FAO and the platform on agrobiodiversity Research.14–16 April 2010, Rome, Italy. Food and Agriculture Organization of the United Nations and the Platform for Agrobiodiversity Research, Rome, p 66

FAO (2019) The state of the World's biodiversity for food and agriculture. In: Belanger J, Pillings D (eds) FAO Commission on genetic resources for food and agriculture assessments. http://www.fao.org/3/CA3129EN/CA3129EN.pdf. Accessed 2 June 2019

Gattuso DJ (2008) Conservation easements: the god, the bad, and the ugly. National Policy Analysis 569, National Policy Center, Washington, DC

Holechek JL (1981) Livestock grazing impacts on public lands. A viewpoint. J Range Manag 34(3):251–254

Holleman H (2017) De-naturalizing ecological disaster: colonialism, racism and the global dust bowl of the 1930s. J Peasant Studies 44(1):234–260. https://doi.org/10.1080/03066150.2016.1195375

Jenkins CN, Van Houtan KS, Pimm SL, Sexton JO (2015) US protected lands mismatch biodiversity priorities. PNAS 112(16):5081–5086

Larson-Praplan S (2018) History of rangeland management. University of California Cooperative Extension. rangelandarchive.ucdavis.edu/Annual_Rangeland_Handbook/History_of_Range_Management/

Mack RN (1981) Invasion of *Bromus tectorum* L. into western North America: an ecological chronicle. Agro-Ecosystems 7(2):145–165

Manheim FT (2010) A brief history of United States forest management policy. In: Dudok V, Lentiakov V, Myklash L (eds) 2010 forestry. Forest and wood technology of the Ukraine, vol 36. National University of Forestry and Wood Technology of the Ukraine, Lviv, Ukraine, p 7

Mooney H, Zavaleta E (2016) Ecosystems of California: threats & responses supplement for decision-making. University of California Press, Oakland, CA 72 p

Morriss A (2007) Europe meets America: property rights in the New World. Foundation for Economic Education, Atlanta, GA

Muhn J, Stuart HR (1988) Opportunity and challenge. The story of BLM. USDI BLM BLM/YA/GI-88/001+1701

Nolon JR (2006) Historical overview of the American land use system. A diagnostic approach to evaluating governmental land use control. Pace Environmental Law Review 23(3):821–852

Oswalt WB, Smith WB (2014) U.S. Forest resource facts and historical trends. USDA Forest Service FS-1035, Washington, DC

Peterken G (2013) Meadows. British Wildlife Publishing, Gillingham, UK

Peters CB, Zhan Y, Schwartz MW, Godoy L, Ballard HL (2017) Trusting land to volunteers: how and why land trusts involve volunteers in ecological monitoring. Biol Conserv 208:48–54

Stubbs M (2014) Conservation Reserve Program (CRP): status and issues. Congressional Research Service, Washington, D.C., pp 7–5700

Thompson JR, Carpenter DN, Cogbill CV, Foster DR (2013) Four centuries of change in northeastern United States forests. PLoS One 8(9):e72540

U.S. Department of Agriculture (2015) Summary report: 2012 National Resources Inventory, Natural Resources Conservation Service, Washington, DC, and Center for Survey Statistics and Methodology, Iowa State University, Ames

US Forest Service (USFS) (2011) Weeks Act centennial: historical legislative and administrative information. https://www.fs.fed.us/land/staff/weeks-act-hist-legislative.html. Accessed 11 June 2019

University of Nebraska, National Drought Mitigation Center (2019) Dust bowl. https://drought.unl.edu/dustbowl/Home.aspx. Accessed 2 June 2019

UNEP WCMC (2018) Protected area profile for United States of America from the World Database of Protected Areas, April 2018. www.protectedplanet.net

USDA Forest Service (USFS) (2019) Forest management. https://www.fs.fed.us/forestmanagement/index.shtml. Accessed 12 June 2019

Vincent CH, Hanson LA, Bjelopera JP (2014) Federal land ownership: overview and data. Congressional Research Service 7–5700, Washington, DC

Whitson TD (1991) Weeds of the west. Western Society of Weed Science, University of Wyoming, Laramie

Warrick RA (1980) Drought in the Great Plains: a case study of research on climate and society in the USA. In: Ausubel J, Biswas AK (eds) Climatic constraints and human activities, IIASA proceedings series, 10. Pergamon Press, New York, pp 93–123

Wishart DJ (ed) (2004) Encyclopedia of the Great Plains. University of Nebraska Press, Lincoln, NE

Chapter 6
Restoring the Future

Ecological restoration is the process by which humans relearn the realities of the place, the importance of acquired wisdom and knowledge, and the relationship that the resident human culture must develop with the place and its biota. (Gerould Wilhelm 1996)

Restoration of degraded ecosystems is a major form of land management that is of increasing importance across the U.S. Links between healthy ecosystems and issues of food, water and energy security, maintenance of biodiversity, and general well-being are recognized at the policy and practical level. Maintaining and restoring ecological resilience helps to buffer the impacts of extreme weather events and provides the land with adaptive potential at a time of global environmental change.

The Society for Ecological Restoration (SER 2004) defines ecological restoration as "*the process of assisting the recovery of an ecosystem that has been degraded, damaged, or destroyed*" (SER 2004; McDonald et al. 2016). Strategies for restoration depend on the severity and duration of ecosystem disturbance and the threatening factors that remain. Where the ecosystem is not too damaged, natural regeneration may take place without assistance as with the return of abandoned farm land to forest vegetation in the eastern states when favorable social, regulatory and ecological conditions are met (Chazdon et al. 2017). At the other end of the scale, where soil has been removed and the land polluted through mining, extensive reconstruction may be required.

It is important to consider the potential for a damaged site to recover naturally, but the option of passive restoration, letting nature take its course, is frequently not realistic. After disturbances caused by road building, fire, clear felling, mining and over grazing, some form of re-stocking with native plants is commonly required. Historically, the treatment of degraded sites has relied on the use of commonly available nursery stock from a limited range of fast growing trees or agriculturally produced seed of forage grasses – often non-native species. Increasingly the limitations of such an approach are being recognized, and the value of an ecological

S. F. Oldfield et al., *Seeds of Restoration Success*, Springer Earth System Sciences, https://doi.org/10.1007/978-3-319-96974-9_6

approach to restoration is widely acknowledged. Developing native plant materials for restoration is challenging but has many advantages for the environment and with the right plant combinations builds in resilience against future disturbances, such as invasions of exotic species and climate change. The use of native plants requires an adequate supply of appropriate seed. Selection and identification of species and seed sources, collection from the wild, storage, seedling production at nurseries, and seed production in agricultural fields are all part of the process enabling this supply and are discussed in this chapter.

Federal agencies are charged with restoration as part of their multi-functional management roles. The USFS is, for example, involved in a cooperative forest restoration program with 23 Collaborative Forest Landscape Restoration (CFLR) sites. The USFS recognizes that the U.S. relies on forests for ecological, social, and economic benefits and that tens of millions of acres of forests and watersheds are in need of restoration so that landscapes will be more resilient to stress. The CFLR sites include different ecosystems, including alpine peaks, pine forests and grasslands. Implementation of the scheme began in 2010, and subsequently CFLR has established more than 67,000 acres (27,000 ha) of forest vegetation, treated more than 1.45 million acres (587,000 ha) to reduce the risk of uncharacteristic wildfire, and improved more than 1.33 million acres (538,230 ha) of wildlife habitat (USFS 2015).

The Bureau of Land Management (BLM) is also actively involved in landscape-level restoration, especially in dryland habitats in the American west. This is in response to a wide range of stresses on the land and is often part of a mandatory mitigation process. The standard definition of mitigation is to take action to avoid, reduce, minimize, rectify, or compensate for adverse impacts (see National Environmental Policy Act 1970). Mitigation is applied as a tool to compensate for the impact of developments, such as mining or road development that damage the environment. It is a very important consideration in multipurpose land management as practiced, for example, by federal agencies. These agencies have developed regional mitigation as a landscape-scale approach to compensate for impacts to resources on federal land. The term mitigation has become closely associated with response to climate change. The Energy Independence and Security Act (2007), for example, mandated an assessment of carbon storage and greenhouse gas fluxes in U.S. ecosystems be undertaken, including an evaluation of potential policies for climate change mitigation.

Ecological restoration is an essential component of mitigation action whether in response to local developments that damage the environment or the global impact of climate change, which is much broader in scope and more difficult to attribute to particular causal agents. To undertake restoration actions, the raw materials – native plants – need to be available. The challenge is to act on a bold and ambitious scale. A major first step, developing native materials for restoration and rehabilitation, is discussed in this chapter.

Box 6.1: Botanic Gardens and Ecological Restoration

Botanic gardens have played a leading role in the conservation of rare and threatened plant species for over 40 years and have built up impressive *ex situ* collections. Increasingly while maintaining this essential role, botanic gardens are now working in partnership with other agencies to support ecological restoration. Botanic gardens have staff with a combination of botanical, horticultural, and ecological skills. They also have unique living collections and seed banks of native species. They have major roles in seed collection, production, storage, research, and practical restoration action either on their own land or in partnership with other land managers. The emphasis in such activities is growing beyond a focus on rare plants to include plants that may be common and needed for ecosystem services.

The Lady Bird Johnson Wildflower Center in Texas, for example, has a research program that examines the effects of fire on natural systems and its use as a tool in restoration. Research results have been applied to the restoration of public land and made widely available to facilitate improved ecological landscape design (LBJWC 2019).

Chicago Botanic Garden's research focuses on ecological restoration and conservation both at its own site and more broadly. Much of their restoration work focuses on arid lands in the western U.S. For instance, in the Colorado Plateau, Garden scientists are compiling data on historical use of species and seed sources for restoration to assess outcomes. They are also identifying and helping develop appropriate native plant materials for restoration that is focused on potential "native winner" species that may improve restoration outcomes in degraded and invaded sites, and quantifying how species and seed-source selection impacts ecosystem function in restored habitats. Closer to Chicago, they also have research programs focusing on comparing ecosystem services provided by intact, degraded, and restored grasslands, woodlands, and wetlands (CBG 2019).

Missouri Botanical Garden is another leading botanic garden active in restoring landscapes both locally and globally. The ecological restoration goal of the Shaw Nature Reserve is the creation or rehabilitation of a mosaic of habitats that can support healthy populations of the maximum possible number of plant and animal species native to the St. Louis region. Approximately half of the Reserve's 2,400 acres (971 ha) are currently set aside as the "Native Habitats Area" with glades, prairies, wetlands, and woodlands. Management methods include assessment of existing diversity, control of exotic species and aggressive natives such as eastern red cedar (*Juniperus virginiana*), reintroduction of native plants and occasionally animals, and properly timed mowing and prescribed burning. Ecological restoration has been underway for over 30 years (MBG 2019).

Sources: CBG (2019), LBJWC (2019), MBG (2019).

Producing Plants for Ecological Restoration

Wild plant reproduction in natural ecosystems ensures regeneration of locally adapted plant species and populations that provide habitat for wildlife, stabilize soils, control surface water flow, and contribute to overall ecological integrity and resilience. Seeds have the ability to renew life and revitalize vegetation. Stored and improved crop seeds have been used in agricultural systems since the beginning of civilization, and over time much of the earth's natural vegetation has been replaced with cropland. Various methods of sophisticated seed storage have been devised, and today, major agricultural seed banks exist to help support the farming industry and guarantee long-term food security. In forestry, seed is collected and stored to restore forests on public lands; seed may be selected to provide new growth of fine quality timber on private lands. Increasingly, with the degradation of natural plant-rich habitats, it is now essential to secure seed of wild plants to save species and their genetic diversity, develop plant materials for ecological restoration and mitigation, support pollinators for ecosystem function and for agricultural production, and to maintain the future of the rich floral heritage of the U.S.

Historically, as noted in Chap. 4, seed banks focused on storing plant materials for food and timber production. In the U.S., recognition of the value of storing native plant materials at a national level for their restoration use dates back only about 20 years. The extensive wildfires of 1999 and 2000 led Congress to direct the BLM and the USFS to facilitate the development of a long-term program to provide a stable and economical supply of native plant materials for restoration and rehabilitation efforts on public lands (House Interior Appropriations Conference Report 106-914 1999-2000). The two federal agencies responded to this by establishing BLM's Native Plant Materials Development Program and USFS's Native Plant Restoration Program. These new developments in natural resource management coincided with global efforts to collect and store the seed of wild rare plants for conservation, coordinated by the Millennium Seed Bank project of the Royal Botanic Gardens, Kew, UK.

The immediate focus of the BLM's Native Plant Materials Development Program was to increase the availability of diverse native plant materials and to foster more efficient management of that supply. Collection and storage of wild-sourced seed enabled a more consistent supply of plant material to provide for variable restoration needs. Foundation material collected from the wild as seed has been given to the private sector who use it to produce more seed – genetically very similar to what came from the wild. Over the past 20 years, the need to scale up native plant production using native seed in support of ecological restoration efforts has become increasingly apparent. The impact of climate change and need for managing ecological resilience are recognized in environmental laws and policies that have a strong requirement for native plant materials (see Appendix 1 for a summary of different laws and policies).

As noted in Chap. 5, the Department of Interior Secretarial Order SO3336 on Rangeland Fire Prevention, Management and Restoration (Jewell 2015) called for enhanced action to prevent and suppress rangeland fire and to restore one particular ecosystem - fire-damaged sagebrush (*Artemisia*) communities that occur extensively throughout the interior West. The Secretarial Order specifically calls for the utilization of locally-adapted seeds and native plant materials for vegetation management and restoration activities.

Other major national policies have been introduced that increase the need for native seed supplies. Advancing the availability and use of pollinator-friendly seed mixes in land management, restoration and rehabilitation is, for example, a specific requirement of the National Strategy to Promote the Health of Honey Bees and Other Pollinators (Pollinator Strategy) published in 2014 (Obama 2014), which requires the restoration or enhancement of 7 million acres (2.833 million ha) of land for pollinators over the next 5 years.

Collection and Storage of Seed

Since 2001, the national initiative known as Seeds of Success (SOS), discussed in Chap. 4, has led the way nationally in making seed collections of the nation's flora. Established by BLM as part of the Native Plant Materials Development Program, and in partnership with several botanic gardens and others, the SOS teams have made more than 24,400 native seed collections covering more than 6,000 taxa. Initially the emphasis was on *ex situ* conservation with the overall goal to bank seed of multiple populations of approximately 14,000 native plant species, around 75 percent of the U.S. native flora. SOS aims to ensure that the native plant seed collections are of high quality, accurately identified, genetically representative of the collected populations and that they are well-documented. The emphasis is now shifting to populations of species having high restoration value; populations of species most at risk from climate change and associated threats; and species representative of key ecological communities. Within the SOS national collection, the most frequently collected species are grasses (Oldfield 2018). Examples are Sandberg's bluegrass (*Poa secunda*), Indian ricegrass (*Achnatherum hymenoides*), squirreltail (*Elymus elymoides*), needle and thread (*Hesperostipa comata*) and basin wildrye (*Leymus cinereus*). Also commonly collected are forbs of the daisy family (Haidet and Olwell 2015).

Specially trained collection teams begin their work as early as January and February in parts of the Southwest, and work continues until late fall in many areas. Locating healthy populations of target plant species is the first step. Populations must then be revisited throughout the field season to ensure collection takes place at the right time in the seed ripening process. Seed collections are cleaned at the U.S. Forest Service's Bend Seed Extractory (BSE) in Oregon. Ten thousand seeds from each collection are sent to the USDA Agricultural Research Service's Western Regional Plant Introduction Station (ARS WRPIS) in Pullman, Washington, where they enter the National Plant Germplasm System (NPGS). The main aim of the NPGS is to safeguard the genetic diversity of agriculturally important plants.

However, ARS also manages seed of native species, including the SOS National Collection. One-third of each ARS sample is placed in storage at 39 °F (4 °C) and small quantities are made available to researchers on request. The remaining two-thirds of each ARS sample is placed in long-term storage (−0.4 °F [−18 °C]) at the ARS National Center for Genetic Resources Preservation in Fort Collins, Colorado. Excess seed from these collections is stored at the BSE and is available for increasing seed production and restoration use. The BSE was established in 1948 to clean, store and propagate conifer seed for reforestation, but it now will clean and store seed of any native species for federal agencies. It forms part of the U.S. Forest Service's nursery network developed to support restoration goals.

Complementing *ex situ* storage of seed in seed banks, the BLM National Seed Warehouse System (BLM Seed Warehouse) ensures that seed is available in the short term for restoration needs. The BLM Seed Warehouse has two major seed warehouses located in Boise, Idaho and in Ely, Nevada. They have a combined capacity of 2.1 million lbs (953,000 kg) of dry storage and 135,000 lbs (61,000 kg) of cold storage. Relying on over 65 major commercial collectors and growers, the BLM Seed Warehouse provides a range of commonly required species such as the ecological workhorse species discussed further below. The BLM Seed Warehouse buys seed three or four times a year and stockpiles it for use after disasters, thus mitigating the effect of demand spikes.

In the eastern U.S., SOS collections are made by several botanic gardens including Chicago Botanic Garden, New England Wildflower Society, North Carolina Botanical Garden, as well as the New York Parks Department. These gardens typically also clean the seed and then either send it to the ARS WRPIS to enter the National Plant Germplasm System with the western SOS material or partner with Plant Materials Centers in the east to propagate material for restoration. Priorities in the east include collections of plant species for tallgrass prairie, coastal dune, and wetland restorations.

Agricultural Seed Production

Most tree and shrub seed is collected from wildland stands or in some cases planted orchards. Many grass and forb species, however, cannot be harvested from the wild in quantities adequate to meet restoration needs. Collected seed of these species must be increased through agricultural field production to provide materials for restoration and related field research. Collecting seed from native species is the first stage in developing adapted seed sources. This may be accomplished to provide a reliable, commercially available source of material for use in large, reasonably homogeneous areas requiring considerable restoration or through contract production for specific projects. Making seed available to commercial growers is very important to produce the required quantity of native seed. Federal plant materials programs have contributed to increased seed production of widespread and common "workhorse" species. These are locally adapted native plants that have broad ecological amplitude, and are generally easy to propagate. Another vital component of the Native Plant Materials Development Program has been the increased seed production by private growers with seed of 122 native species becoming commercially available and seed of nearly 200 additional species in development. Germination and propagation protocols have been developed for over 3,000 native species (BLM 2009).

In general, for large scale restoration programs, parent seed of commonly used grasses and forbs is increased in agricultural fields maintained by private growers, federal agencies, and universities and then distributed to growers for commercial production. For woody shrub species it is generally considered uneconomical to grow plants under agricultural conditions because of the time required before the first seed crop can be harvested. In the U.S. there are at least 841 vendors of native plant species (White et al. 2018). Nevertheless, the range of native plants available still remains relatively limited in species, seed source, and quantity for restoration purposes. In the Western states, for example there are around 14,600 native plant taxa. A recent study has shown that only 1,949 of these native plants are currently used in commercial seed production, usually at small scale (White et al. 2015). Nationally 74 percent of the U.S. vascular plant flora is unavailable for restoration unless material is sourced from the wild (White et al. 2018). This severely impacts restoration potential for natural areas that have been degraded.

Growers generally remain reluctant to begin production of a wider range of native seed in an uncertain and often fluctuating market (Shaw and Jensen 2014). Efforts to diversify and scale up native seed production depend to a large extent on adequate and reliable funding for development and ultimately on market demand. Federal agencies are major buyers of native seed for land rehabilitation and restoration. BLM, the largest federal land manager, alone purchases 3 million pounds (1.4 million kg) of seed per year and needs much more to meet restoration requirements.

Box 6.2: The Work of the Plant Materials Centers
The USDA Natural Resources Conservation Service (NRCS) operates 25 Plant Materials Centers (PMCs), each based in ecologically distinct areas, to evaluate plants and vegetative technologies to support USDA conservation programs and practices. PMCs help to find vegetative solutions to reduce soil erosion, increase soil health and productivity in croplands, improve water and air quality, improve habitat for wildlife, restore wetlands, and stabilize coastal areas. The PMCs cooperate with many public and private conservation partners to collect, evaluate, select, and release plants for commercial production to support resource conservation.

Source: NRCS (2019)

Selection of Native Plant Species for Restoration

Deciding which plant species to use in restoration can be complex. Usually the intention is to replicate the species composition at a particular site based on knowledge of what grew there previously or is growing nearby. But recreating nature is not easy. For practical purposes, a small subset of plants may be chosen initially to kick-start the ecological recovery process. In restoration programs, important plants are the workhorse species. They are the first species selected for planting and will help to provide a sense of place as well as enabling the establishment of other species. It is important to identify these common plants for each region as they can perform well in degraded sites and potentially facilitate succession to diverse native plant communities.

Another important consideration in selecting species for restoration projects is the interaction of native plants with exotic invasive plant species. Many of these exotics are adapted to disturbance conditions, so actual and potential invasions must be considered when planning restoration projects. Management to reduce invasion potential includes alleviation of disturbance regimes, reducing propagule pressure of the exotics, and site preparation practices to remove existing exotics and their seed banks. An additional tool is careful selection of revegetation species to provide plants capable of establishing and developing rapidly that will capture available resources (nutrients, light, water), particularly those that most favor the exotics in question. Planting practices such as water catchment or use of seedlings to enhance establishment of slower-growing species may be required. Reestablishment of a diverse, functional community resistant to invasion is the goal, but ongoing monitoring and management of weed populations may be required.

To encourage the use of regionally appropriate plants in both landscaping and ecological restoration, preliminary lists of restoration-ready plant species have been prepared for Level III ecoregions (see Fig. 6.1), and in some cases, particular species recommendations have been developed for restoration of different habitats within the ecoregions. Practitioners can use this information in conjunction with analyses of conditions at restoration sites and species identified in local reference areas when developing seeding and planting strategies. Lists are available from a number of sources. Ecoregional lists of species for roadside revegetation are available on the Ecoregional Revegetation Application (U.S. DOT FHA 2018) and recommendations for pollinator species on the USDA NRCS Ecoregional Planting Guide website (USDA NRCS 2018).

For all species selected, ecologists consider it desirable to use native plant material sourced locally and adapted to the restoration area. Species populations evolve in their local habitats adapting to climate, soil conditions, and the effects of other plants and animals. Using locally adapted plant material improves the likelihood of restoration success and helps to maintain the local natural variation among plants and associated animals. As mentioned above, recent national policy directives on fire management, pollinators and climate change all call for the use of locally adapted native plant species. Deciding how local is appropriate is a key factor in successful restoration.

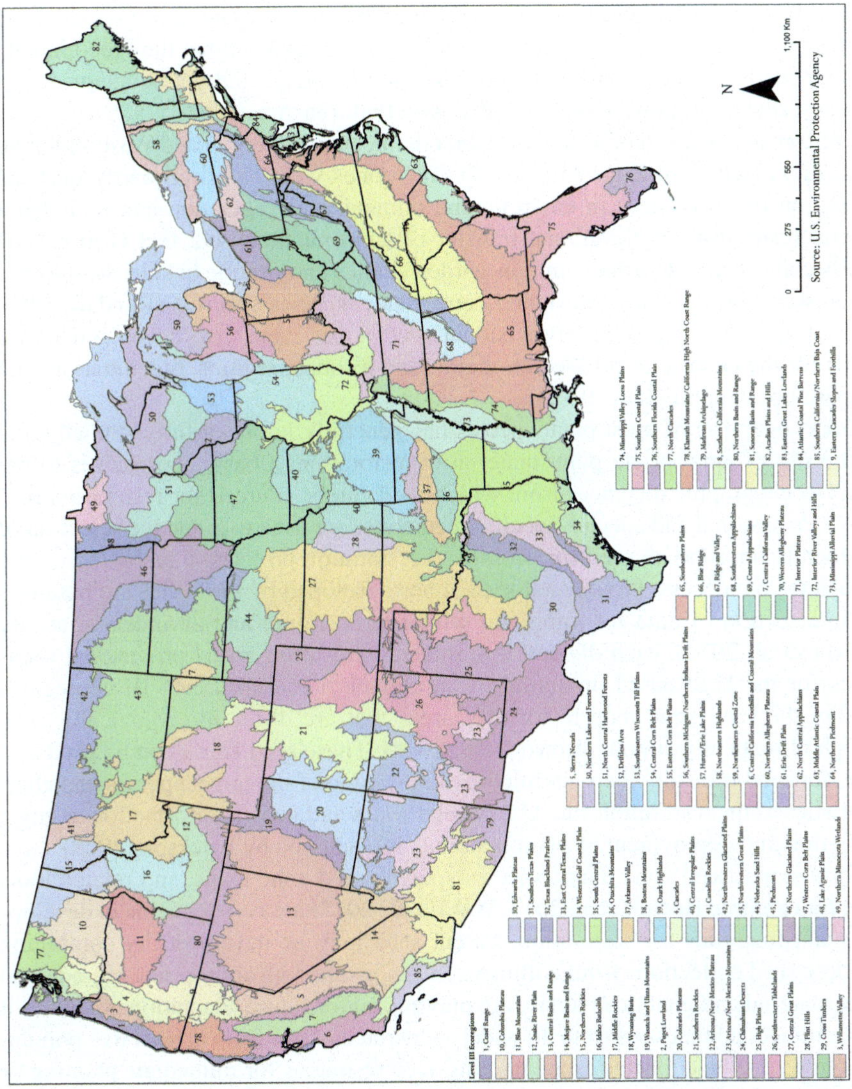

Fig. 6.1 Omernik Level III Ecoregions (Omernik 1987)

Development of Seed Zones

In selecting plant material of appropriate provenance for restoration, seed zones are used. A seed zone is a mapped area with fixed boundaries in which seeds or plant materials can be transferred with the expectation that they are appropriately adapted. Seed zones are widely used in forestry and are beginning to be used in restoration of other native plants. Seed transfer guidelines are similar in that they recommend how far seeds can be transferred from the point of origin and yet be adapted, and they describe the relative risk associated with that transfer.

For forest species, seed zones were first delineated in Oregon and Washington 50 years ago (Tree Seed Zone Map 1973). The zones were based primarily on local evaluation of differences in environment, climate, and vegetation and included a general restriction on elevational transfer of approximately 492 feet (150 m). As data became available from common-garden studies and genetic field trials, the seed zones were broadened, redesigned, and made species-specific (Randall 1996, Randall and Berrang 2002). Provisional seed zones were also developed in the 1970s for the Great Plains based on major land resource regions, precipitation, and temperature (Cunningham 1975).

Local adaption of most plant species has generally not been fully studied so to ensure use of appropriate plant material in restoration, a basic approach is to put limits to geographic distances from which seed can be sourced for restoration purposes. Provisional and species-specific seed zones are used to guide the selection of genetically appropriate seed for ecological restoration projects, thereby increasing the chance of success. Provisional seed zones developed by the USFS are based on the intersection of data for minimum winter temperature and an index of aridity (Bower et al. 2014), each divided into intervals, resulting in 64 provisional seeds zones for the U.S. based on similarity in climate. Omernik Level III ecoregions overlay the seed zone map to identify areas that are different ecologically even though climatically similar (Bower et al. 2014). These seed zones can be used as a basis for seed transfer when modified by knowledge of the target species and characteristics of the restoration site. Use of seed zones and genetic information to guide collection and deployment of plant materials is discussed by Erickson (2008).

Provisional seed zone maps, species-specific seed zone maps, and related tools (see, for example Howell et al. 2017, WWETAC 2017) are very valuable in developing plant materials for restoration. Parent seed may be developed by combining multiple seed collections from within a zone to ensure both appropriate pre-adapted plant material and genetic diversity are maximized for future adaptation potential. It is recommended that seed from 50 widely separated plants from at least five populations in a zone be collected when pooling seed resources for immediate planting or seed increase for native plant development (Johnson et al. 2010).

The provenancing strategies described above are further complicated by climate change. As our climate continues to change, species may be marooned at sites where they are no longer adapted. Some restoration practitioners are already adjusting sourcing guidelines (i.e., using seed from more distant provenances), often with climate change in mind. Types of provenancing approaches that consider climate change include: *relaxed local provenancing* (Kaye 2001), *composite provenancing* (Broadhurst et al. 2008), *admixture provenancing* (Breed et al. 2013), *predictive provenancing* (e.g., Crowe and Parker 2008), and *climate adjusted provenancing* (Prober et al. 2015). These approaches need further empirical testing across a broad array of species and locations before best practices can be recommended. Some species may be able to migrate effectively or adapt to new conditions. In addition, the restoration context and the species can affect restoration sourcing decisions. A conservative sourcing approach may be warranted for narrow endemics with little long distance gene flow and for species that occur in very stable, high-quality, and unique habitats (Havens et al. 2015).

Native Plants for Restoration of Forests

In the U.S., there is a long history of using native conifer species in reforestation programs, but the use of a wider range of native plants in restoration programs is relatively new. It is now national policy of the USFS to use native plant materials as first choice when timely natural regeneration of the native plant community will not occur (USFS 2012). In the Pacific Northwest a strategic plan has been developed to ensure that suitable and affordable native plant materials are available and utilized (Erickson 2008). Initial selection of priority species and quantities required across U.S. Forest Service land in the Pacific Northwest (over 25 million acres [10.1 million ha]) involved assessment of vegetation at reference sites and restoration areas together with review of literature, vegetation and soil maps, plant association guides, and information on fire occurrence. Prioritization of species also took into account the feasibility of collection of plant material from the wild and potential for production of seed or seedlings for restoration. Initially identified workhorse species included a number of important grasses and woody plants such as willows (*Salix* spp.) and cottonwoods (*Populus* spp.).

Following the identification of USFS priority species in the Pacific Northwest, plant material procurement plans are developed for each species. For common species that are readily produced, commercial growers generally produce the plant materials required. Collection from the wild is also an important part of procurement involving USFS staff together with volunteers and professional collectors. Guidelines are used to ensure representative sampling of genetic variation. At least 50 percent of the seed crop at a collection site is left in place to ensure natural recruitment and regeneration of the native population. Wild collected seeds are sent to the BSE in Oregon. In addition to wild collection, native materials for restoration are increased through seed or seedling production in USFS nurseries in Oregon and Idaho. Traditionally these nurseries grew known-provenance, locally adapted conifer seedlings for reforestation, but their role has broadened over the past 20 years to meet the needs of ecological restoration. Additional public and private facilities are now taking on this task to meet an expanding market. Germination and propagation protocols for Pacific Northwest species are developed by the Corvallis PMC in Oregon working closely with the USFS.

In the Southeastern U.S., major efforts are underway to restore the longleaf pine (*Pinus palustris*) forests that have been severely depleted by timber and turpentine exploitation, conversion to agriculture, and alteration of fire regimes (Brockway et al. 2006). The Longleaf Alliance, and other private and public partners, have created a plan to expand the area of longleaf forests to 8 million acres (3.2 million ha) by 2024. Longleaf pine still occurs over most of its natural range, although in isolated fragments within varied vegetation types, and so habitat restoration on a large scale is feasible. Restoring natural authenticity is considered a reasonable goal, to restore compositional, structural, and functional components of the forests. One of the limitations in producing the quantities of longleaf pine seedlings needed for restoration has been the lack of sufficient high-quality seeds. Guidelines to address this issue are provided by Barnett and McGilvray (2002). Various methodologies for restoration are being used based on the habitat type with fire management in all cases being an essential component.

Box 6.3: Restoration of the Appalachian Forests

The rich temperate forests of the Appalachians, introduced in Chap. 3, have declined as a consequence of logging and coal mining. Coal mining has created a need for forest restoration in extensive areas of the region. One of the problems is that mining completely disturbs the soils so that fertile top soil in which trees flourish is buried under layers of unproductive material. In early restoration attempts, native trees failed to thrive, whereas invasive species established, forming shrublands with limited biodiversity value. The Appalachian Regional Reforestation Initiative (ARRI) is a partnership-based program, established in 2004, with the aim of restoring former mine lands. The partnership has developed guidelines for restoring seriously disturbed lands to native forest. The Forestry Reclamation Approach provides five steps to successful reforestation including how to get the soil "right." The use of both ecologically important early successional tree species and commercial crop tree species is promoted. With funding from the Appalachian Regional Commission, an organization called Green Forest Works is re-establishing healthy and productive forests on formerly mined lands in the region. More than 9,000 Green Forest Works volunteers have planted more than one million trees on nearly 2,000 acres (809 ha) of former mined land since 2009 (OSMRE 2019). In addition, along with other organizations, Green Forest Works is reforesting 1,000 acres (447 ha) of reclaimed mine land with American chestnut (*Castanea dentata*), a native species that was found throughout the Appalachians until it was devastated by a fungal disease in the early years of the last century. The American Chestnut Society, which has donated trees for use in recovering mined land, U.S. Forest Service scientists and many others have developed a strain of American chestnut tree that tolerates the disease that nearly wiped out the species a century ago.

Source: OSMRE (2019)

Native Plants for Restoration of Prairies

The demand for restoration of prairie vegetation is increasing, but as with other ecosystem types, appropriate seed resources are limited. Locally sourced and grown seed has been shown to outperform non-local seed in restorations, but there are relatively few nurseries specializing in source-identified local seed. In the Chicago region, for example, a recent survey found that most nurseries do not have seed available from local sources (within 50 miles [80.5 km] or less) for many plant species, so that restoration practitioners are adjusting their seed-sourcing criteria to within 100-200 miles (160-3201 km) or more of the city (Havens, unpub. data).

In Iowa, only an estimated 0.01 percent of the original tallgrass prairie still remains. Much of the rich, fertile land has been converted to corn and soybean production. The almost complete destruction of the ecosystem has led to an interest in using Iowa-sourced seeds for increase and commercial production for restoring the remnant tallgrass prairie. The Tallgrass Prairie Center at the University of Northern Iowa has several programs aimed at restoration on the few remaining areas available for perennial vegetation in an agricultural landscape: roadside rights-of-way, prairie strips within row crop agricultural fields, and indirectly on federally subsidized Conservation Reserve Program (CRP) pollinator plantings. Currently, Iowa farmers and landowners have about 1.6 million acres (650,000 hectares) enrolled in the federal CRP, much of it planted to prairie, with a recent emphasis on forb-rich mixes for pollinator recovery. Recent work by Iowa State University's STRIPS team has demonstrated that planting prairie in strips on as little as 10 percent of row crop acres can mitigate up to 90 percent of sediment and nutrient run-off from agricultural practices. Prairie strips can stop erosion, reduce nutrient loss, improve soil and water quality, and support pollinators and other wildlife.

For the past 27 years, Natural Selections, formerly the Iowa Ecotype Project, has collected and increased seed from remnant prairies throughout Iowa as source-identified material for seed increase. Greenhouse grown seedling plugs are transplanted into seed increase nursery beds for three provenance zones – southern, central, and northern Iowa. Certified as Iowa Source-Identified seed, in accordance with Association of Seed Certifying Agencies (AOSCA) guidelines, seed is released and further increased and certified by commercial seed producers in and around Iowa. Since 1990, the project has made over 3,000 original collections of seed, and released nearly 400 lots of seed of over 60 species to 17 growers for commercial production. Over 1.25 million pounds (570,000 kg) of seed have been produced from these releases since 1996. In addition to Natural Selections seed releases, private growers can collect and increase their own source material. The Iowa Crop Improvement Association annually certifies Source-Identified seed of 140-180 native species (Houseal 2017). Some Iowa native plants remain hard to find for prairie restoration purposes including several grasses, sedges, and a number of forbs that are difficult to grow or germinate.

Box 6.4: Restoration of Oak Savannas

Oak (*Quercus* spp.) savannas are now rare ecosystems scattered within the Great Plains that are considered to be globally imperiled. There is, however, potential to restore oak savannas, particularly in hill country where agriculture is less dominant. The Driftless Area of the Midwest, which constitutes a large, mostly hilly area in southwestern Wisconsin with small outliers in Illinois, Iowa, and Minnesota, probably has the largest amount of restorable oak savanna in the Midwest. Various Wisconsin State Natural Areas are being managed for oak savanna. Also, savannas are being managed and restored in Minnesota, Iowa, northern Illinois and Indiana, southwestern Michigan, and northwestern Ohio. Farther west, a large number of savannas are under management in Missouri. In Illinois, oak savanna restoration has become an important activity in a suburban band around the Chicago metropolitan area. The Chicago Wilderness Oak Recovery Program, established in 2012, has involved mapping oak ecosystems, explaining their significance to broad audiences, managing oak savanna areas, and promoting the planting of appropriate species. Also, extensive restoration work is under way by The Nature Conservancy and others in the Oak Openings Region of northwestern Ohio. Indiana Dunes National Lakeshore is undertaking oak savanna restoration at that location.

The importance of oak savanna restoration has been recognized by the USDA Natural Resources Conservation Service (NRCS) through its Wildlife Habitat Incentives Program (WHIP), and special "points" are given for evaluating WHIP proposals for private landowners that involve significant savanna habitat. The State of Wisconsin has promoted oak savanna restoration by private landowners through its Landowner Incentive Program. The oak savanna has also been singled out for concern by state agencies of most of the Midwestern states that have significant restorable savannas (Oak Savannas 2019).

Sources: Darling and Fahey (2016), Oak Savnnas (2019)

Native Plants for Restoration of Deserts

The vegetation of the cold deserts of the U.S. has declined in area and quality as a consequence of the spread of introduced species, increased drought resulting from global climate change, wildfires, and other disturbances. The fragmentation of sagebrush habitats has, for example, resulted from a range of threats, exacerbated by the spread of invasive weeds, such as cheatgrass (*Bromus tectorum*), which fuels more intense wildfires. Federal and state agencies are making major efforts to restore sagebrush habitats on their lands. In January 2015, an Order by the Secretary of the

Department of the Interior called for enhanced action to prevent and suppress rangeland fire and to restore fire-damaged, sagebrush communities (Secretarial Order 3336, Jewell 2015). One of the purposes of the Order was to prevent the decline of the greater sage-grouse (*Centrocercus urophasianus*), an iconic species and game bird, closely associated with sagebrush landscapes that has declined dramatically because of loss of habitat.

Greater sage-grouse are dependent on sagebrush as described in Box 3.3. Sagebrushes are keystone species enabling the successful germination of other forbs and grasses, which help to support pollinators, grazing wildlife, and domestic livestock. Restoration of sagebrush habitats following burning has traditionally involved seeding with plant materials developed for forage production. The emphasis now is on increased use of native plants and development of new or modified seeding techniques to accommodate multiple seed types. Action to manage and restore sagebrush helped to ensure that listing of the greater sage-grouse under the Endangered Species Act was not considered necessary following review and extensive consultation in 2015. Additionally, Secretarial Order 3336 specifically calls for the utilization of locally adapted seeds and native plant materials for vegetation management and restoration.

Restoration is being planned or implemented on a major scale and requires huge quantities of native plant species. Broad partnerships between federal and state agencies, non-governmental conservation organizations, universities, and the private sector are helping to make available the plants required. The Great Basin Native Plant Project, led by the USFS and BLM, was the first such partnership, established in 2002, as a response to the impact of major wildfires (Shaw et al. 2012). The Great Basin has about half of the remaining sagebrush communities harboring the greater sage-grouse and over 200 other species of conservation concern. The goal of the Project is to increase seed availability and develop methodologies to restore native plant communities across millions of acres of burned lands with a focus on restoring native sagebrush habitat and increasing native forb production.

A critically important aspect of research is understanding the genetic diversity of the widespread and abundant plant species to ensure availability of adapted plant materials for restoration of particular sites. Given the importance of sagebrush in the Great Basin, reestablishment of adapted species of appropriate genetic provenance has been a major component of the Great Basin Native Plant Project. In years of intensive burning, land management agencies may purchase over 500,000 pounds (227,000 kg) of seed for this region. Unfortunately, there have been problems in misidentification and mixing of Wyoming big sagebrush (*Artemisia tridentata* ssp. *wyomingensis*) and basin big sagebrush (*A. t.* ssp. *tridentata*) by seed collectors, potentially leading to less successful restoration outcomes. Techniques for long-term storage of sagebrush seed have been developed together with techniques for growing and developing nursery stock (Karrfalt and Shaw 2013, Dumroese et al. 2015) increasing its restoration potential.

The Great Basin Native Plant Project has evaluated 91 genera in addition to *Artemisia* with a total of 225 native plants, about 80 percent of which are forbs. This represents about 15 percent of the regional flora (Dumroese et al. 2015). Examples of species of importance for restoration that remain in short supply are given in Box 6.5.

Box 6.5: Great Basin Priority Forbs for Restoration

The following native species are important for restoration but remain in short supply. All have been produced successfully in seed fields. Research has addressed the earlier obstacles to production of most of these species, though some challenges remain. Collection of appropriate seed sources and commercial production are often required.

Arrowleaf balsamroot (*Balsamorhiza sagittata*): An excellent species for pollinators and other wildlife that can be grown in seed fields, but flowering does not begin until about the third year, so it is generally not considered an economically viable seed crop. Seed is collected from large stands that exist in the wild. However, crops are erratic as wild seed production is often impacted by late frost, insect predation, and browsing by deer or livestock.

Basalt milkvetch (*Astragalus filipes*): A legume that is beneficial for soil fertility.

Two Select class releases are produced in seed fields and available in limited amounts.

Dusty maiden (*Chaenactis douglasii*): An annual or short-lived perennial that flowers in the first year and can be valuable for pollinators in the year following wildfires.

Gooseberryleaf globemallow (*Sphaeralcea grossulariifolia*): Provides fodder for wildlife and pollen and nectar for bees. This species is available erratically in moderate amounts with occasional production in seed fields. Commercially available seed is harvested from wildland stands in the southern Great Basin, but stands in the northern Great Basin are generally smaller and do not yield enough seed for most projects.

Hoary aster (*Machaeranthera canescens*): An early successional, short-lived perennial that develops rapidly and flowers in autumn. It is a useful species for disturbed sites and post-fire seedings.

Limestone hawksbeard (*Crepis intermedia*): An important species for sage-grouse, but it remains difficult to produce.

Nineleaf biscuitroot (*Lomatium triternatum*): A valuable source of spring forage for wildlife, and it has some competitive ability with cheatgrass. It is long-lived, but generally does not produce a good seed crop until the third year. Growers are producing Source-Identified germplasms for the Northern Great Basin Ecoregion.

Royal penstemon (*Penstemon speciosus*): A species that occurs in Wyoming big sagebrush and disturbed sites within its range. This species is useful for restoration of disturbed sites such as roadcuts.

Shaggy fleabane (*Erigeron pumilus*): A widespread species that could be included in many seed mixes as it is good for wildlife and pollinators.

Sulphur-flower buckwheat (*Eriogonum umbellatum*): A very widespread, generally long-lived species that is useful for adding diversity to wildlands and color for native landscape projects.

Silky lupine *(Lupinus sericeus)*: A nitrogen-fixing species, important pollen source for bumble bees, nectar source for honey bees, and favored by hummingbirds. Establishment and insect damage to seeds are common production problems.

To the south of the Great Basin, native ecosystems of the Colorado Plateau have also been severely impacted. Colorado Plateau vegetation consists of pinyon-juniper woodland communities at higher elevations with saltbush-greasewood (*Atriplex-Sarcobatus* spp.) and blackbrush (*Coleogyne ramosissima*) communities at lower elevations. In the past, degraded areas have been planted with introduced grasses but there is now a shift toward ecological restoration using native plant species. The Colorado Plateau Native Plant Program (CPNPP) was created by the BLM in response to a lack of sufficient variety and quantity of genetically appropriate native plant materials for ecological restoration in the Colorado Plateau ecoregion (Wood et al. 2015). The working partnership includes federal and state agencies, nonprofit organizations, and universities.

An initial step in this Program was to develop a list of priority species required for restoration. A working list of 100 species was developed that was then refined by an expert committee, taking into account the potential for agronomic increase together with provision of ecosystem services. A working list of 33 priority forbs and 11 grasses was produced. As a result of partnership programs, seven grass species and six forb species have been developed for commercial production and will be made available to private sector growers for increased market availability of local native plant materials (Wood et al. 2015).

It remains a major challenge for the CPNPP to capture ecotypic diversity within abundant species that need to be produced commercially to support restoration efforts. In support of this program, the U.S. Geological Service (USGS) and partners are conducting population genetic studies on selected Colorado Plateau species. Working with the National Park Service (NPS), the group is using genetic markers to evaluate the spatial scale of genetic differentiation in Indian ricegrass, sand dropseed, and smallflower globemallow (*Sphaeralcea parvifolia*) collected from locations across Arches National Park, Canyonlands National Park, and adjacent BLM lands in southeastern Utah. The NPS is ultimately interested in using these species to restore degraded grasslands within the parks (Wood et al. 2015).

In a larger-scale study, the USGS is genotyping populations of blue grama grass (*Bouteloua gracilis*) drawn from across the Colorado Plateau. Sampling is focused to evaluate the impact of precipitation timing and elevation, both of which vary strongly across the Plateau. In addition, population sampling of the forbs, scarlet gilia (*Ipomopsis aggregata*) and showy goldeneye (*Heliomeris multiflora*), is underway.

The PMCs in Meeker, Colorado; Los Lunas, New Mexico; and Aberdeen, Idaho provide information on native plant materials they have made available through intensive research and development. In addition, the Uncompahgre Partnership has been evaluating several native species collected from the Colorado Plateau. Through their efforts, native plant materials are becoming available for commercial growers throughout the West.

The development of native plant materials for restoration is also underway in the Mojave Desert, with the BLM leading the ecoregional approach. One focus for restoration has been the habitat of the Mojave desert tortoise (*Gopherus agassizii*), a species listed as Threatened under the Endangered Species Act. Seventy-four percent of the designated Critical Habitat for this species occurs on land managed by the BLM. Large areas of suitable land have been set aside in Nevada to conserve the Mojave desert tortoise with grazing allotments retired to reduce competition for forage. The Mojave Desert equivalent to cheatgrass in the Great Basin is red brome (*Bromus rubens*), an exotic invasive annual grass that fuels fires that damage native vegetation and seed banks and increases soil erosion. Red brome in the diet has been shown to reduce growth and survival in juvenile Mojave desert tortoises. To combat these problems, habitat managers are attempting to restore thousands of acres of degraded critical habitat to enhance the recovery of the Mojave desert tortoise. The diet of the Mojave desert tortoise consists primarily of annual forbs and grasses and many short-lived perennials. Food plants important for Mojave desert tortoise habitat management include: desert globemallow (*Sphaeralcea ambigua*), desert dandelion (*Malacothrix glabrata*), ovate plantain (*Plantago ovata*), Booth's suncup (*Eremothera boothii*) and desert wishbone bush (*Mirabilis laevis*). Cover and habitat stabilization plants that are important include creosotebush (*Larrea tridentata*) and jointfir (*Ephedra* spp.). Most of these plants can be acquired commercially, but the seed sources of these materials are generally only broadly known. Current research, sponsored by BLM, is helping to develop a wider range of plants for restoration taking into account the need for locally adapted material.

Native Plants for Restoration of Coastal Wetlands

In late October 2012, Hurricane Sandy impacted a wide swath of the East Coast of the U.S., resulting in more than 120 deaths and major disaster declarations for 12 states plus the District of Columbia. The Hurricane Sandy Disaster Relief Supplemental Appropriations Act, 2013 specifically directed the use of mitigation funds for restoration of natural areas and for increasing coastal habitat resilience.

As a consequence, the U.S. Fish and Wildlife Service (FWS) is working with partners on 31 projects designed to increase resilience by restoring coastal marshes, beaches, and dunes; improve connectivity in streams and rivers; and foster collaborative science. One of the projects is, for example, restoring and enhancing 432 acres (175 ha) of damaged coastal salt marshes on the east coast of Long Island. Many of the restoration projects require a supply of native plants, but these have not been readily available.

The Mid-Atlantic Regional Seed Bank (MARSB), established in 2011 by New York City's Greenbelt Native Plant Center, collects seeds, trains collectors, and educates users on the need for genetically appropriate plant material. To address the lack of plant materials for post-Sandy restoration, MARSB works with partners including BLM, Chicago Botanic Garden (CBG), New England Wild Flower Society (NEWFS), North Carolina Botanic Garden (NCBG), and Cape May Plant Materials Center (CMPMC) to collect appropriate seed from throughout the region to support FWS, NPS, and other coastal restoration projects. A list of over 40 essential workhorse species has been developed as a basis for providing the essential plants for ecological restoration of coastal ecosystems. Additional species are required to ensure local variation. Over the course of the project, 2,167 seed collections were made for 356 species found in sub-tidal habitats, beaches and dunes, wetlands, coastal forests, and inland rivers and streams. More species will be added to develop a long-term conservation resource in the form of banked seed available for future resource management and restoration.

Box 6.6: Building Ecological Resilience in New York

In response to Hurricane Sandy, the Governor's Office of Storm Recovery (GOSR) was created to coordinate the recovery efforts of New York State and invest its allocation of $4.4 billion in Community Development Block Grant – Disaster Recovery funding from the U.S. Department of Housing and Urban Development (HUD). GOSR aims to strike a balance between urgent action to address rebuilding needs and strategic planning to build resiliency solutions for the long-term. The New York State recovery programs recognize that sea-level rise and increasingly frequent instances of extreme weather events highlight the need to protect and restore coastal ecosystems. Strategic retreat programs, like the State's Buyout Program, are designed to reflect the ecology of the areas they relate to as well as to the communities they serve. Operating in select coastal and riverine neighborhoods, the Buyout Program purchases properties of interested homeowners, whose homes were substantially damaged or destroyed. The Buyout Program improves the resiliency of the larger community by transforming properties into open space or natural stormwater management systems, creating a coast buffer to safeguard against the impacts of future storms.

Source: GOSR (2019)

Native plants for ecological restoration in New York City are produced by the Greenbelt Native Plant Center. This facility also provides a home for the Mid-Atlantic Regional Seed Bank (MARSB). The mission of the Greenbelt Native Plant Center is to propagate and grow native plants from appropriate local ecotypes for New York City habitat restoration. Seed is collected from local plant populations and used to grow plants of appropriate size. The Center propagates over 336 species of native plants from local genetic source material that is best adapted to conditions in the area to ensure long-term sustainability. Seed collecting techniques are designed to capture maximum genetic diversity by: collecting seed from a minimum of 50 individuals from any given plant population; recognizing ecotypic differences; respecting isolation distance requirements for source-identified seed certification; and harvesting early, mid, and late maturing seed. Currently forty species are under production in the foundation seed program including three species of goldenrod (*Solidago*) and three species of aster (*Symphyotrichum*).

National Action: Supporting Pollinators

Implementation of the National Pollinator Strategy requires increased availability of flowering plants chosen to bloom at suitable times to provide nectar and pollen for different species of insects and other pollinators. The Strategy requires the development, storage, and use of native plant materials to provide increased food for pollinators as well as nesting, cover, and forage habitat. The Pollinator Enhancement Plan specified that the highest priority native plant species for pollinators be identified for all ecoregions. NRCS Plant Materials Centers are working with Xerces Society (https://xerces.org), State and Federal agencies, local native plant societies, the seed industry, and other partners to create or update NRCS Technical Notes that outline pollinator-friendly species by ecoregion. (https://plants.usda.gov/pollinators/NRCSdocuments.html).

Sowing seed for pollinators in restoration or rehabilitation projects benefits from the use of plant species that have long flowering seasons and attract multiple pollinators, as well as supporting other wildlife. Getting the right mix of flowering plants is also important with the emphasis on including groups of species that flower at different times through the season. Currently, plants that bloom early in the season are missing from many pollinator seed mixes (Havens and Vitt 2016). A mix of species with different life forms, including pioneering and perennial species should be considered. The Pollinator Partnership, a non-profit organization, has produced Ecoregional Planting Guides (https://www.pollinator.org/guides) to show land managers, farmers and gardeners across the country a selection of plant species that are appropriate for attracting and supporting pollinators in their region. The plant species included are generally available in trade and are suitable for a variety of conditions. Using the guides will provide a tremendous boost for pollinators. At the same time, however, there is an urgent need for more research into plant-pollinator relationships and for action to increase the availability and range of pollinator-friendly plant species.

Box 6.7: Wild Bees' Needs and Restoration

In North America, especially for the prairies and sagebrush-steppe habitats, skilled commercial seed growers are producing seed of diverse native grasses and forbs. Farmed wildflower seed is cheaper; production is more plentiful and reliable; and the resulting seed is often of better quality than wild-harvested seed—all market factors necessary for widespread adoption and largescale application. The Great Basin wildflowers in line for restoration efforts are all widespread species. They represent diverse plant families, and happen to be both attractive to native bees and are in need of bees' pollination services. One-third of all mason bee species (Osmia spp.), Figure 6.2, of North America were found visiting the flowers of basalt milkvetch (*Astragalus filipes*), a Great Basin forb in commercial seed production. This is also true for many widespread prairie forbs useful for restoration in the Great Plains. For example, purple prairie-clover (*Dalea purpurea*) is in commercial seed production on several hundred acres in five states. It both requires bees for seed set (a critical factor for growers) and is used by 50+ species of native bees for pollen and nectar. A close prairie relative, leadplant (*Amorpha canescens*), retains a rich bee fauna, including a dominant specialist, even among small prairie remnants. This additional evidence indicates that relict bee communities are poised to take advantage of nearby revegetation efforts. Recent advances in seed growing, harvest, cleaning, storage, dormancy management, and especially mechanical seeding technologies are facilitating successful restoration demonstrations. Research and field trials coordinated through the Great Basin Native Plant Project and funded by BLM Plant Conservation & Restoration Program and the USFS are at the vanguard of such rehabilitation programs. Native wildflower restoration is the key to bee community restoration in all habitats.

Source: Cane (2011)

Fig. 6.2 Female bee of Osmia texana. Photo by Aaman Mengis, USGS Bee Inventory and Monitoring Lab, Laurel MD

National Action: The National Seed Strategy

All the native seed development initiatives discussed above are reliant on partnerships. At a national level it is important to harmonize efforts to plan for and develop native seeds for ambitious restoration goals. The National Seed Strategy for Rehabilitation and Restoration 2015-2020 (PCA 2015) has been developed to provide a more coordinated approach for the stabilization, rehabilitation and restoration of ecosystems through the provision of appropriate native seed. For effective implementation of the Strategy, it is crucial for federal and state agencies and non-governmental organizations to work with the private sector to develop native seed to meet restoration needs. at a time when the impacts of extreme weather events are increasing. The Strategy recognizes the importance of healthy native plant communities as an essential foundation for ecosystem integrity and diversity, providing ecosystem services that sustain people, their communities, their health, and their livelihoods.

Implementation of the Strategy will lead to the coordinated establishment of national networks of native seed collectors, seed researchers, seed growers, restoration scientists, and seed users. The Strategy builds on a range of city initiatives, such as NYC's Greenbelt Native Seed Nursery or Chicago Wilderness; state initiatives such as Iowa's Tallgrass Prairie Center; and regional initiatives, such as those of the Southeastern Grasslands Initiative, Great Basin, Pacific Northwest, and Longleaf Alliance, scaled up to ensure adaptation and ecological preparedness in response to climate change. It provides a framework to support national ecological restoration policy objectives. More broadly, it will help to support food security by incorporating wild crop relatives into storage and restoration plans and will support water security by providing appropriate plants for restoring water catchment areas and landscapes experiencing increased drought.

The National Seed Strategy builds on the seed collecting work undertaken by the BLM and USFS over the past 20 years. The four inter-related goals of the Strategy are:

Goal 1: Identify seed needs and ensure the reliable availability of genetically appropriate seed reserves.

Goal 2: Identify research needs and conduct research to provide genetically appropriate seed reserves and to improve technology for seed production and ecological restoration.

Goal 3: Develop tools that enable managers to make timely, informed seeding decisions for ecological restoration

Goal 4: Develop strategies for internal and external communication.

The challenges of collecting and developing native plant materials remain considerable but achievable with a recent paper (Havens et al. 2015) offering the following recommendations:

1. Prioritize action on essential or workhorse species that provide a native matrix for ongoing restoration. These may be early successional native species. Develop regional lists of these species and bulk up large quantities of their seed with the source of the material identified.
2. Collect and bank genetically diverse seed samples of as many other native plant species as possible to provide a full complement of botanical diversity for restoration and an insurance policy against climate change.
3. Increase short-term (cool or warehouse conditions) and long-term (frozen) seed storage capacity for workhorse species and the additional range of native plants appropriate for different ecoregions.

With the launch of the National Seed Strategy, land managing agencies, organizations, and private individuals have a stake in its success and need to work together to facilitate the development and use of native plants in restoring degraded ecosystems and native plant communities across the U.S. A business plan was developed to determine the cost of implementing the Strategy and highlight the opportunities for the private sector to become more involved in the multi-million dollar native seed industry. Non-governmental organizations are actively involved in the Strategy. The 2017 National Native Seed Conference, with over 330 delegates from 41 states and 11 nations, highlighted the multitude of actions already underway in native seed production at local, state, and ecoregional levels and the need to be fully coordinated. The 2019 National Seed Strategy Progress Report features work conducted to address the Strategy's four goals. It also reports on ecoregional projects that show the extent of collaborations underway to create a more comprehensive network of collectors, testers, and growers (BLM 2019).

Native seed and plant materials are being used in innovative ways to celebrate the botanical diversity throughout the country. Using native plants along the extensive U.S. road network provides connectivity across an increasingly fragmented landscape for plants, wildlife, and pollinators. Highways cross virtually every ecosystem in the country, and roadside vegetation often serves as a pathway for exotic plant invasion. Replacing invasive weeds with appropriate native plant species helps establish corridors for gene flow between isolated ecosystems and provides contiguous habitat for plants and wildlife. By selecting native plants with appropriate growth habits, maintenance requirements are reduced at the same time as providing an aesthetically pleasing flora that reflects the local native plant communities of the area.

The magnificent U.S. landscape with its diverse ecosystems and vegetation cannot be returned to pristine condition. But with an understanding of botanical diversity and its relationship to ecological resilience we can make intelligent and informed choices for the future, as we discuss further in the next chapter.

References

Barnett JP, McGilvray JM (2002) Guidelines for producing quality longleaf pine seeds. General technical report SRS–52. USDA Forest Service, Southern Research Station, Asheville, NC, 21 p

Bureau of Land Management (BLM) (2009) Native plant materials development program. Progress Report for FY2001-2007 Washington, DC

Bureau of Land Management (BLM) (2019) Bureau of Land Management Native Plant Communities. www.blm.gov/seedstrategy. Access 12 June 2019

Bower AD, St. Clair JB, Erickson V (2014) Generalized provisional seed zones for native plants. Ecol Appl 24(5):913–919. https://doi.org/10.1890/13-0285.1

Breed MF, Steed MG, Otewell KM, Gardner MG, Lowe AJ (2013) Which provenance and where? Seed sourcing strategies for revegetation in a changing environment. Conserv Genet 14:1–10

Broadhurst LM, Lowe AJ, Coates DJ, Cunningham SA, McDonald M, Vesk PA, Yates CJ (2008) Seed supply for broadscale restoration: maximizing evolutionary potential. Evol Appl 1:587–597

Brockway DJ, Outcalt KW, Tomczak DJ, Johnson EE (2006) Restoration of longleaf pine ecosystems. General Technical Report SRS-83. USDA Forest Service, Southern Research Station, Asheville, NC, 34 p

Cane JH (2011) Meeting wild bees' needs on Western US rangelands. Rangelands 33(3):27–32

Chazdon RL, Bodin B, Guariguata M, Lamb D, Walder B, Chokkalingam U, Shono K (2017) Partnering with nature: the case for natural regeneration in forest and landscape restoration. FERI Policy Brief, Montreal, 11 p

Chicago Botanic Garden (CBG) (2019) Chicago botanic garden. https://www.chicagobotanic.org/. Accessed 11 June 2019

Crowe KA, Parker WH (2008) Using portfolio theory to guide reforestation and restoration under climate change scenarios. Clim Chang 89:355–370

Cunningham RA (1975) Provisional tree and shrub seed zones for the Great Plains. Research paper RM-150, USFS Rocky Mountain Forest and Range Experiment Station, Fort Collins, CO, 15 p

Darling L, Fahey RT (2016) Sustaining oaks in the Chicago region landscape: developing a plan for maintaining oak dominance in an urban landscape. Int Oaks 27:195–206

Dumroese RK, Luna T, Richardson BA, Kilkenny FK, Runyon JB (2015) Conserving and restoring habitat for greater sage-grouse and other sagebrush-obligate wildlife: the crucial link of forbs and sagebrush diversity. Native Plants J 16(3):277–295

Energy Independence and Security Act of 2007. Pub. L.110-140. 121 Stat. 1492. (December 19, 2007). https://www.congress.gov/110/plaws/publ140/PLAW-110publ140.pdf. Accessed 17 Jan 2018

Erickson VJ (2008) Developing native plant germplasm for national forests and grasslands in the Pacific Northwest. Native Plants J 9(3):255–266

Haidet M, Olwell P (2015) Seeds of Success: a national seed banking program working to achieve long-term conservation goals. Nat Areas J 35(1):165–173

Havens K, Vitt P (2016) The importance of phenological diversity in seed mixes for pollinator restoration. Nat Areas J 36:500–550

Havens K, Vitt P, Still S, Kramer AT, Fant JB, Schatz K (2015) Seed sourcing for restoration in an era of climate change. Nat Areas J 35(1):122–133

House Interior Appropriations Conference Report 106-914 (1999-2000) Making appropriations for the Department of the Interior and related agencies for the fiscal year ending September 30, 2001, and for other purposes. https://www.congress.gov/congressional-report/106th-congress/house-report/914. Accessed 18 Apr 2018

Houseal G (2017) Twenty-seven years of source-identified seed developments, production and prairie restoration on Iowa's roadsides and agricultural lands. http://ser-insr.org/news/2016/10/27/twenty-seven-years-of-source-identified-seed-development-production-and-prairie-restoration-in-iowas-roadsides-and-agricultural-lands. Accessed 18 April 2018

Howe G, St. Clair JB, Bachelet D (2017) Seedlot Selection Tool. USDA Forest Service, Oregon State University and Conservation Biology Institute, Corvallis, OR, https://seedlotselection-tool.org/sst/. Assessed 27 Apr 2018

Jewell S (2015) Secretarial Order Number 3336. Rangeland fire prevention, management, and restoration. U.S. Department of the Interior, Washington, DC. https://www.forestsandrangelands.gov/rangeland/documents/SecretarialOrder3336.pdf. Accessed 18 Jan 2018

Johnson R, Stritch L, Olwell P, Lambert S, Horning ME, Cronn R (2010) What are the best seed sources for ecosystem restoration on BLM and USFS lands? Native Plants J 11(2):117–131

Karrfalt R, Shaw N (2013) Banking Wyoming big sagebrush seeds. Native Plants J 14(1):60–69

Kaye TN (2001) Common ground and controversy in native plant restoration: the SOMS debate, source distance, plant selections, and a restoration-oriented definition of native. In: Rose R, Haase D (eds) Native plant propagation and restoration strategies. Nursery Technology Cooperative and Western Forestry and Conservation Association, Corvallis, OR, pp 5–12 Reprinted in: Grasslands (2007) 17:4-9

Lady Bird Johnson Wildflower Center (LBJWC) (2019) Lady Bird Johnson Wildflower Center. https://wildflower.org/. Accessed 11 June 2019

McDonald T, Gann GD, Jonson J, Dixon KW (2016) International standards for the practice of ecological restoration – including principles and key concepts, 1st edn. Society for Ecological Restoration, Washington, D.C., 47 p

Missouri Botanical Garden (MBG) (2019) Missouri botanical garden. https://www.missouribotanical garden.org/ Accessed 11 June 2019

New York State Governor's Office of Storm Recovery (GOSR) (2019) Governor's Office of Storm Recovery. https://stormrecovery.ny.gov/. Accessed 11 June 2019

Oak Savannas (2019) Oak Savannas. https://www.oaksavannas.org. Accessed 19 June 2019

Obama B (2014) Presidential memorandum—Creating a federal strategy to promote the health of honey bees and other pollinators. Washington, D.C.: The White House. https://obamawhitehouse.archives.gov/the-press-office/2014/06/20/presidential-memorandum-creating-federal-strategy-promote-health-honey-b. Accessed 18 Jan 2018

Office of Surface Mining Reclamation and Enforcement (OSMRE) (2019) Appalachian regional reforestation initiative. https://arri.osmre.gov/About/AboutARRI.shtm. Accessed 11 June 2019

Oldfield S (2018) The National Seed Strategy for rehabilitation and restoration: progress and prospects. Plant Biol 21(3):380–382

Plant Conservation Alliance (PCA) (2015) National seed strategy for rehabilitation and restoration 2015-2020. https://www.blm.gov/seedstrategy Accessed 27 Apr 2018

Prober SM, Byrne M, McLean EH, Steane DA, Potts BM, Vaillancourt RE, Stock WD (2015) Climate-adjusted provenancing: a strategy for climate-adaptable ecological restoration. Front Ecol Evol 3:65

Randall WK (1996) Forest tree seed zones for western Oregon. Oregon Department of Forestry, Salem, p 82

Randall WK, Berrang P (2002) Washington tree seed transfer zones, vol 63. Washington Department of Natural Resources, Olympia

Shaw N, Jensen S (2014) The challenge of using native plant materials for sagebrush steppe in the Great Basin, USA. In: Kiehl K, Kirmer A, Shaw N, Tischew S (eds) Guidelines for native seed production and grassland restoration. Cambridge Scholars Publishing, Newcastle upon Tyne, pp 141–159

Shaw N, Pellant M, Fisk M, Denney E (2012) A collaborative program to provide native plant materials for the Great Basin. Rangelands 34:11–16

Society for Ecological Restoration International Science and Policy Working Group (2004) The SER International primer on ecological restoration. Version 2. Society for Ecological Restoration International, Tucson, AZ, 13 p. http://c.ymcdn.com/sites/www.ser.org/resource/resmgr/custompages/publications/SER_Primer/ser_primer.pdf. Accessed 18 Apr 2018

The National Environmental Policy Act of 1969, as amended (Pub. L. 91-190, 42 U.S.C. 4321-4347, January 1, 1970, as amended by Pub. L. 94-52, July 3, 1975, Pub. L. 94-83, August 9, 1975, and Pub. L. 97-258, § 4(b), Sept. 13, 1982) https://ceq.doe.gov/. Accessed 17 Jan 2018

USDA Natural Resources Conservation Service (NRCS) (2018) Plants Database. Fact sheets and plant guides. Washington, DC, https://plants.usda.gov/java/factSheet. Accessed 18 Apr 2018

US Department of Transportation, Federal Highway Administration (US DOT FHA) (2018) Ecological Revegetation Application. Washington, DC, http://www.nativerevegetation.org/era/. Accessed 2 June 2019

USDA Forest Service (USFS) (2012) Native plant materials policy: A strategic framework. Washington, DC. 15p https://www.fs.fed.us/wildflowers/Native_Plant_Materials/documents/NativePlantMaterialsPolicy_Sept2012.pdf. Accessed 27 Apr 2018

USDA Forest Service (USFS) (2015) Collaborative forest landscape restoration program 5-year report. USDA Forest Service, Washington, DC. FS-1047. 16 p. https://www.fs.fed.us/sites/default/files/2015-march-cflr-fs1047.pdf. Accessed 17 Jan 2018

USDA Forest Service, Western Wildland Environmental Threat Assessment Center [USFS WWETAC] (2017) TRM Seed Zone Applications. https://www.fs.fed.us/wwetac/threat-map/TRMSeedZoneMapper.php. Accessed 13 Oct 2017

USDA Forest Service and Western Forest Tree Seed Council (1973) Tree seed zone map. USDA Forest Service in cooperation with the Western Forest Tree Seed Council, Portland, OR

USDA Natural ResourcesConservation Service (NRCS) (2019) USDA Natural Resources Conservation Service Plant Materials Program, https://www.nrcs.usda.gov/wps/portal/nrcs/main/plantmaterials/pmc/. Accessed 11 June 2019

White A, Fant J, Kramer A (2015) Restoring species diversity: are vulnerable plant species falling through the cracks? Paper presented at the National Native Seed Conference; 13–16 Apr 2015, Santa Fe, NM

White A, Fant JB, Havens K, Skinner M, Kramer AT (2018) Restoring species diversity: assessing capacity in the United States native plant industry. Restor Ecol 26(4):605–611

Wilhelm G (1996) What is ecological restoration and why is it important? Conservation design forum. https://www.cdfinc.com/xm_client/client_documents/Importance_of_Ecological_Restoration.pdf

Wood TE, Doherty K, Padgett W (2015) Development of native plant materials for restoration and rehabilitation of Colorado Plateau ecosystems. Nat Areas J 35(1):134–150

Chapter 7
Looking Ahead

"The total value of biodiversity is infinite, so having a debate about what is the total value of nature is actually pointless because we can't live without it." (Robert Scholes in Salles 2011)

Plant diversity is essential for our future. Many actions underway, including those described in Chaps. 4, 5, and 6 of this book, show that plant diversity in the U.S. can be effectively conserved and restored. Learning from these successes, initiatives need to be increased. The U.S., as elsewhere in the world, is experiencing a period of unprecedented environmental change with implications for national and global economic development and security. The World Economic Forum lists extreme weather, water shortages, natural disasters, and a failure to prepare for climate change as four of the greatest perils in terms of impact in its *Global Risks Report* 2017, only surpassed by weapons of mass destruction. There are major challenges in managing native plant diversity, including "plant blindness" that leads to a lack of general public and political interest in plants or plant conservation. Botanical capacity is declining at the same time as it is necessary to study and conserve native plant communities and understand the shifting baselines arising from climate change. There is a race against time to develop and implement species recovery plans, protect biodiverse land, collect and store native seed, and establish reintroduction protocols. In this chapter we outline the challenges and highlight the innovative and collaborative approaches to maintain plant diversity and the ecological resilience it provides to ensure that options are secured for future generations.

In recent years, various reviews (for example, Mooney 2010, Ellis 2011, Brook et al. 2012, Ellis et al. 2013) have documented the extent to which the growing demands of humanity outstrip natural capacity to replenish and renew and have therefore become unsustainable. The rapid pace of habitat loss, the growing number of endangered species, the accelerating extinction rate, and the increasing speed of global climate change have been well documented in the summary of the IPBES Global Assessment Report on Biodiversity and Ecosystem Services (Diaz et al. 2019). The biodiversity crisis is real. In the U.S. there are a number of examples of species and community recovery resulting from a myriad of collaborative

S. F. Oldfield et al., *Seeds of Restoration Success*, Springer Earth System Sciences, https://doi.org/10.1007/978-3-319-96974-9_7

conservation actions. Conservation successes need to be recorded and celebrated. The National Seed Strategy, introduced in Chap. 6, provides a remarkable opportunity to do so for native plants of the U.S.

Box 7.1: The UN Sustainable Development Goals

The UN Sustainable Development Goals (SDGs) set out global aspirations for the integration of economic development, natural resource management and protection, and social equity and inclusion. The U.S. is working to implement the Sustainable Development Goals (SDGs), both domestically and overseas. Adopted in 2015, the SDGs explicitly recognize the fundamental and inseparable interdependence between humanity and nature. There are 17 SDGs covering all aspects of human life and well-being. SDG 15 aims to protect, restore, and promote sustainable use of terrestrial ecosystems, sustainably manage forests, combat desertification, halt and reverse land degradation, and halt biodiversity loss. Other goals rely on biodiverse ecosystems for their achievement including SDG 2, which focuses on food security and sustainable agriculture, and SDG 11, which aims to make cities and human settlements inclusive, safe, resilient, and sustainable. Within the U.S., five cities (New York City, Minneapolis, New Orleans, San Jose, Santa Monica) and one state (California), have committed to achieve the SDGs, and companies such as Johnson and Johnson and Coca-Cola acknowledge their importance (UN 2019). As President Obama stated in 2015.

> We commit ourselves to new Sustainable Development Goals...we recognize that our most basic bond—our common humanity—compels us to act...we reaffirm that supporting development is not charity, but is instead one of the smartest investments we can make in our own future.

Sources: Obama (2015), UN (2019)

Progress in Ecological Restoration

Positive biodiversity conservation achievements over recent years include the progress in collection and long-term storage of wild seed and the increasing recognition of the importance of ecological restoration. Globally over 10 percent of wild plants have their seed in safe storage; in the U.S. the figure is closer to 25 percent. Examples of ecological restoration initiatives are numerous, and the need for increased restoration has been noted in a range of international policy documents, including the GSPC introduced in Chap. 1, the UN Biodiversity Targets, the UN Decade on Ecological Restoration, and the UN SDGs (see Box 7.1.). Worldwide, governments have committed in the Bonn Challenge to restoring 1.4 million square miles (3.5 million km^2) of degraded landscapes by 2030.

Box 7.2: Demonstrating 80 Years of Prairie Restoration Success

As described in Chaps. 3 and 5, the iconic prairies of the U.S. have suffered major degradation over the past 200 years and are now subject to increasing restoration by private and public bodies. Commercial nurseries and landscaping companies specialize in the planting of prairies and/or selling prairie plants or seeds to homeowners and private corporations that use native prairie plants for landscaping. Prairie grasses and forbs are planted along major highways in various states.

Curtis Prairie, of the University of Wisconsin Arboretum, represented the first major attempt at prairie restoration and has provided inspiration for over 80 years. It is the site of pioneering restoration work carried out by Aldo Leopold and other celebrated ecologists. The area of Wisconsin Arboretum was previously oak (Quercus) savanna, settled by Europeans in 1837, and Curtis Prairie was formerly a horse pasture. Curtis Prairie was purchased by the University for research in 1932/1933. Fassett and Thomson began the restoration project in 1935, experimenting with various methods to establish native plants, whether by seed or transplanting sod. Leopold and others worked with the Civilian Conservation Corps to add more species and reduce exotic species. The use of prescribed fire was introduced on an experimental basis in the late 1930s and continues to be used as a management tool. John Curtis, for whom the prairie is named, was instrumental in setting up long-term monitoring of the restoration, which continues to the present day. Invasive species remain a constant threat, thus the restoration requires, and will continue to need, ongoing management. Nevertheless, over the past several decades the Curtis Prairie has provided a chance for millions of visitors to see what the tallgrass prairie once looked like, and this educational role is a critical part of its mission. The Curtis Prairie is widely cited as the oldest restored prairie with research, practical action, and public education demonstrating what is possible with dedicated effort.

Source: Havens (2017)

In the U.S., the USFS has committed, through the Bonn Challenge, to restore 37 million acres (15 million ha) by 2020. Working in partnership with government and state agencies, tribes, non-governmental organizations, and private landowners, this commitment aims to create more climate resilient ecosystems, restore watersheds, increase biodiversity, reduce hazardous fuels, and produce forest products. Investment in large-scale restoration is helping to maintain water supplies in a cost effective way by prioritizing watersheds as well supporting local communities. Reporting and monitoring progress is undertaken through an interdisciplinary, collaborative approach.

It can be clearly demonstrated that ecological restoration not only improves ecological resilience but also brings positive economic benefits. The ecological restoration sector in the U.S. directly employs about 126,000 workers and generates approximately $9.5 billion in economic output (sales) annually. This activity supports an additional 95,000 jobs and $15 billion in economic output through indirect (business-to-business) linkages and increased household spending (BenDor et al. 2015). An analysis of a wide variety of ecosystem restoration projects on federal lands has also quantified the economic benefits. Specifically, a study by Haefele et al. (2016) indicated that ecosystem restoration projects provide meaningful economic contributions to local, regional, and the national economy, with between $2.2 and $3.4 million in total economic output contributed to the U.S. economy for every $1 million invested in ecosystem restoration. Investments in ecosystem restoration support jobs and livelihoods, small businesses, and rural economies.

Over the past 30 years, ecosystem markets in the U.S. have been developed to connect landscape conservation and restoration initiatives with people or organizations willing to pay for these activities, whether to safeguard ecological values that they depend on, to meet regulatory compliance, or to provide a public good. Market activity has grown rapidly to more than 2,400 initiatives across the U.S., driven by factors such as enabling policy, development activity that can trigger demand for ecosystem credits, and the existence of protected or threatened ecosystem types (Bennett et al. 2016). Typically, markets focus on a single ecosystem service or asset type, including wetland and stream conservation, carbon sequestration, conservation of threatened species and habitats, and watershed services. Many factors drive growth in ecosystem markets. California, Minnesota, the Mid-Atlantic states, Georgia, and Florida are areas with a high number of projects in development.

The Need for Seed

Looking ahead, increasing ecological restoration will continue to require technical and scientific challenges to be addressed, critical monitoring of progress to be undertaken, and the needs of local people to be incorporated into restoration projects. A fundamental requirement is to ensure that adequate amounts of native seed and seedlings are available for restoration projects. Despite major progress in seed-collecting over the past two decades in the U.S., a shortage of genetically-appropriate seed of native plant species continues to be a limiting factor. Seed is a critical natural resource that has been largely unrecognized, unprotected, and undermanaged. Appropriately adapted seed sources are critical for restoration and management because they do not pose a genetic risk to surrounding native populations. However, U.S. seed resources are in danger not only from extirpation but also from misuse when maladapted seed is used. As such, national seed resources are in need of the same kind of forward-thinking management as is required for other natural resources such as timber and oil.

Forward-thinking management of the seed resource involves appropriate seed sourcing for restoration under climate change. The "local seed is best" concept has long guided ecological restoration and there are many examples of local adaptation and "home-site advantage" in plants (Leimu and Fischer 2008, Hereford 2009, Baughman et al. 2019). Practitioners now understand that local adaptation varies by species, population, and habitat and a "local" genotype may occur over wider areas (i.e. from 10s to 100s of miles) depending on the species and its biology. This means that, while the local gene pool still has a major role in adaptation, it is prudent to consider including at least a small amount of germplasm of the same species from a future climate – that is, a region with a climate similar to that which is predicted for the area being restored. The inclusion of "pre-adapted" plant material sourced from regions more similar to what the future climate is likely to be is discussed, for example, by Sgro` et al. (2011).

More experimentation is needed to understand the consequences of different seed sourcing approaches. It is clear, however, that reintroducing genetically diverse, regionally sourced plant materials remains a good management practice, particularly when sourced from environmentally similar locations (Broadhurst et al. 2008). It is also clear that seed zones may need to become more dynamic in the future (Kramer and Havens 2009). Tools and data are needed to provide guidance at the population level. The seedlot selection tool (SST) is one such example. This is a GIS mapping program, developed by USFS, Oregon State University, and the Conservation Biology Institute, designed to help restoration practitioners match seedlots with planting sites based on climatic information.

The U.S. National Seed Strategy is tackling the national seed shortage and addressing the issue of genetically appropriate plant material at a time of rapid climate change. It acknowledges the major technical and scientific challenges and capacity needs. The Strategy recognizes that success on a national scale will be achieved through coordinated establishment of a nationwide network of native seed collectors, a network of farmers and growers working to develop seed, a network of nurseries and seed storage facilities to supply adequate quantities of appropriate seed, and a network of restoration ecologists who know how to put the right seed in the right place at the right time.

Achieving successful outcomes of the National Seed Strategy will require additional investments, research, improved decision tools, and enhanced communication. But most importantly, successful implementation of this Strategy will require the active participation of a diverse set of public and private partners. Increased coordination is vital to accelerate the pace and scale of restoration and provide native plant materials when and where they are needed. To that end, the Strategy helps coordinate the efforts of local, state and federal government agencies, Tribes, and the private sector to restore plant communities that have been altered by natural or human-caused events. Full implementation of the National Seed Strategy will potentially stimulate rural economies by helping native seed farmers expand their customer base and increase their seed output.

Coordination to achieve the outcomes of the National Seed Strategy is being demonstrated at all levels as people see the linkages between plant diversity, ecological health, and resilience. At a state level, important conversations in Colorado provide an example of new thinking. The State of Colorado experienced record wildfires and floods in 2012 and 2013 that inflicted nearly $5 billion in damages to housing, infrastructure, and the environment and changed the lives of thousands of Coloradans. Recognizing the need to reduce risk from future disasters, as well as proactively plan for rapid growth and changing conditions, the state developed the Colorado Resiliency Framework, adopted by Gov. John Hickenlooper in 2015. This provides a call to action and partnership, seeking to empower those with a stake in resiliency to participate in making it an ongoing priority and reality. Links with the National Seed Strategy are being actively explored.

Box 7.3: Rehabilitating People and the Land

Restoring the land on a large scale requires a dedicated workforce. An innovative and cost-effective approach developed in six western states (see Fig. 7.1) provides meaningful employment for inmates, as well as a skill for them after prison. The Great Basin Sagebrush Project is part of the Sustainability in Prisons initiative, an environmental partnership between the Institute for Applied Ecology (IAE), Departments of Corrections, and the BLM. Since 2014, more than 1.5 million sagebrush and other sage steppe plants were produced by incarcerated men and women in 12 prisons in 6 different states. Inmate crews, staff, and a horticultural educator assist BLM in planting sagebrush each fall or early winter. Teamwork, mutual respect, and a stewardship ethic among individuals are encouraged. In the words of one inmate.

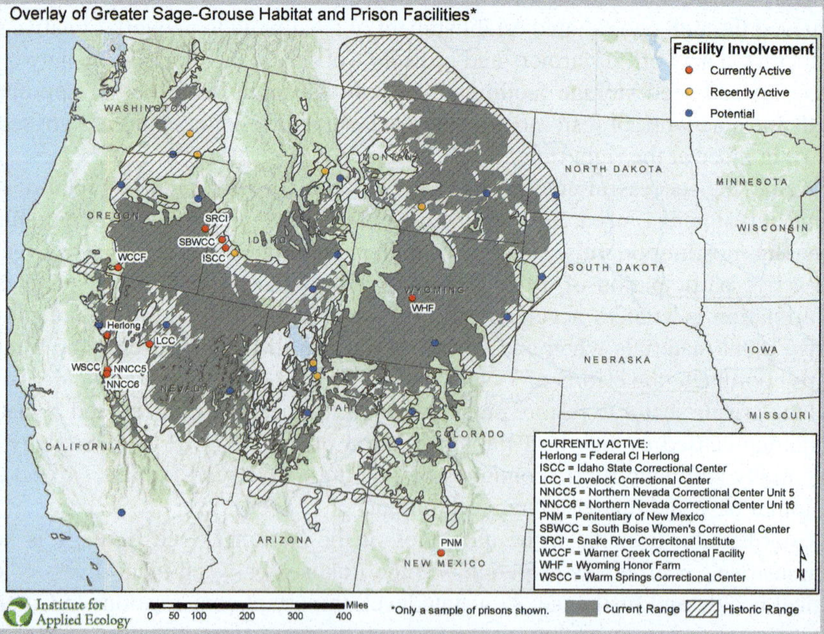

Fig. 7.1 Overlay of Greater Sage-Grouse Habitat and Prison Facilities

> The sagebrush program has been good for my soul. This project has done more than rescue the sage grouse, it has allowed us men to do and be something positive. Through this project, men's' lives begin to have worth; we see that we can be positive and benefit the world instead of bringing it down. Someone smiling and saying well done can move a man's heart and give him back hope. Yes, I've made mistakes, we all have, but the one I don't want to make is missing the chance to give back to the world that has taken care of me.
>
> Source: IAE (2019)

Overcoming Challenges

Full implementation of the National Seed Strategy will require addressing botanical capacity in the U.S. A recent survey documented severe shortages of management and research staff with botanical degrees, suggesting government agencies currently lack the botanical skills required to guide effective management of the nation's critical biological resources (Kramer et al. 2013). Botanical skill shortages were shown to occur throughout all federal and state government agencies, with some of the most significant found in agencies directly responsible for managing federal lands. Furthermore, botanical expertise at federal agencies will continue to decline over the next 15 years as more than half of the current workforce retires. Unfortunately, the decline in botanical infrastructure at government agencies is occurring at the same time as declines in botanical education and training opportunities at U.S. universities. The private sector, including botanic gardens and environmental NGOs, is filling some of the gaps in botanical education and research. More partnerships and training opportunities among academic, private, and government sectors are required to deal with future needs in basic botanical knowledge, invasive species control, climate change adaptation and mitigation, habitat restoration, and the preservation of ecosystem services.

Implementation of the National Seed Strategy will also require financial investment. Fundamental plant science is currently underresourced, receiving just 2 percent of extramural spending for life sciences research in the U.S. (McCormick and Tjian 2010). This is particularly unfortunate, given the economic benefits of plants for food, medicines, construction, energy, and provision of the national green infrastucture as outlined in Chap. 2. The relative disregard for plants in funding terms is also reflected in budgets for conservation of species under threat. Despite plants comprising the majority of the federal endangered species list (57 percent), in 2011 they received less than 4 percent of federal endangered species expenditures. If state and federal expenditures are totaled, plants receive less than 4 percent of the funding for endangered species nationwide (Havens et al. 2014). As of 2019, spending patterns have not changed significantly.

However, there are reasons to be cautiously optimistic. People love nature! A recent estimate of the total economic value of the National Park Service (NPS) to the American public is $92 billion. Two-thirds of this total ($62 billion) is for National Park lands, waters, and historic sites; the remaining $30 billion is attributed to NPS programs. The estimate, which is based on very conservative assumptions, includes not only the value attributed by visitors to the parks, but also a significant "non-use" or "existence" value. This is the value derived by the public from simply knowing that NPS assets are protected for current and future generations, regardless of whether or not they actually choose to visit (Haefele et al. 2016). Furthermore, the ten U.S. organizations leading the environmental movement collectively have 15 million members and an annual budget of more than $525 million (Bagley 2014).

Overcoming Plant Blindness

The term plant blindness has been defined broadly by Wandersee and Schussler (1998) as "the inability to see or notice the plants in one's own environment, leading to the inability to recognize the importance of plants in the biosphere and in human affairs." They noted that plant blindness also comprises an "inability to appreciate the aesthetic and unique biological features" of plants and "the misguided, anthropocentric ranking of plants as inferior to animals, leading to the erroneous conclusion that they are unworthy of human consideration."

Overcoming plant blindness is one of the tasks taken on by botanic gardens. The kind of activities that can be used to engage people with plants include practical gardening to emphasize diverse functions of plants, creating art to promote close observation, exploration of plant movement, and highlighting the human-like traits of individual plant species (Balding and Williams 2016). The work of botanic gardens in conservation and ecological restoration has been described in Chaps. 4 and 5. Complementing these activities, another vitally important role is in public outreach and education. Visitors to major botanic gardens in the U.S. amount to over 41.7 million per year. The United States Botanic Garden (USBG) in Washington, D.C. was the first botanic garden to be established in the nation and is rooted in the country's heritage. During the late eighteenth century, George Washington, Thomas Jefferson, and James Madison shared the dream of a national botanic garden. In 1816, the Columbian Institute for the Promotion of Arts and Sciences in Washington, D.C. proposed the creation of a botanic garden to collect, grow, and distribute plants of the U.S. and other countries that might contribute to the welfare of the American people. The Institute's garden was established by Congress in 1820 to the west of the Capitol grounds, from First Street to Third Street between Pennsylvania and Maryland Avenues. Now, the mission of USBG is to demonstrate the aesthetic, cultural, economic, therapeutic, and ecological importance of plants to the well-being of humankind.

As shown in Fig. 4.2, there are currently over 400 botanic gardens and arboreta in the U.S. Gardens, such as Chicago Botanic Garden, are leading the way in connecting people with plants from cradle to grave and are reaching out to all sectors of society. In order to broaden the appeal of its work, the board and staff of Chicago

Botanic Garden sought to ensure that the garden is relevant to its audiences, while at the same time being a convener of people of different backgrounds, bringing about positive social impacts. As noted by Shaw and Schwarz-Ballard (2017), in describing their approach to engagement of the Chicago Botanic Garden.

> Botanic gardens can achieve operational sustainability and build understanding of plant conservation only if everyone—from diverse neighborhoods and backgrounds—has the opportunity to become involved in our institutions. Let's celebrate the fact that our living museums have the capacity, because of our commitment to nature, to address a diverse suite of needs; both those of our planet and of its diverse and complex population.

Brooklyn Botanic Garden is another garden noted for its outreach and education. The chief mission of the Garden is to educate the public about plants and, as an extension, inform people about ecology and awareness of the environment. This is achieved by organizing "safaris" to the Steinhardt Conservatory, the greenhouse complex with plants from around the world, running practical workshops, and conducting a range of projects aimed particularly at children. Brooklyn Botanic Garden was the first in the world to develop a Children's Garden, inaugurated in 1914. Project Green Reach, which has been running for nearly 30 years, aims to encourage the study of botany and environmental science in poorer urban communities of New York (Brooklyn Botanic Garden 2019).

In addition to botanic gardens and other non-governmental organizations, federal land managing agencies have recognized the importance of educating the general public through various environmental education and outreach programs. Specifically, BLM and the Institute for Applied Ecology developed three regionally-based curricula for grades 9–12. These curricula introduce students to botany and the connection between plants and their ecosystems (www.appliedeco.org). Celebrating Wildflowers is the USFS public outreach project for educating the public about the many values of native plants (https://www.fs.fed.us/wildflowers/). Reaching out in innovative new ways and engaging the wider public in understanding the fundamental importance of plants may help to overcome "plant blindness."

Box 7.4: Budburst

Budburst is a national citizen science initiative launched in 2007 that engages the public in observation and recording of phenological events (the timing of leafing, flowering, and fruiting) for selected wild and cultivated species in the U.S. The goals of Budburst are to (1) increase awareness of phenology as an area of scientific study, (2) increase awareness of the impacts of changing climates on plants, and (3) increase science literacy by engaging participants in the scientific process.

Budburst has engaged participants of all ages and backgrounds through its on-line educational and data-entry program, providing the opportunity for anyone to actively participate in scientific research. Project data are freely available for anyone to download. Budburst is owned and managed by the Chicago Botanic Garden.

Source: CBG (2019)

One particular way to foster engagement is through citizen science. This is a means of expanding research capacity as well as providing stimulating opportunities for participants to be directly involved in conservation research and management (Ellwood et al. 2017). Phenological studies of plants (recording the timing of leafing, flowering and fruiting) have become an effective citizen science activity as described in Box 7.4. The collection of ecological data is another important citizen science activity. Volunteer-based monitoring of land trusts in the U.S. has become an increasingly important way for land trusts to manage their land and engage with the communities within which they operate. In 2010, land trusts had over 136,000 active volunteers. A recent study has shown that plants are the most common monitoring target (Peters et al. 2017).

Box 7.5: The Landscape Profession Taking Inspiration from the Wild

Demand for native plants is increasing across the U.S. This demand results from federal, state, and community ordinances that require or recommend use of native plant species, increases in the number of landscape restoration and reclamation projects, increases in the number of cities and communities with water restrictions for landscaping, and increased awareness of pollinator issues. The increased demand is thought to be enhanced by the expanded use of native plants by landscape architects and contractors. Naturalistic planting has become fashionable. There is increasing recognition that designing with plant communities both links nature to landscapes and brings together ecological planting and traditional horticulture. The idea of designed plant communities provides more diversity and improved ecological function while recognizing that landscapes, especially in urban areas, no longer resemble historically natural conditions. In ecological restoration projects, material developed from known wild collections is required, whereas in urban situations cultivars of native plants are generally acceptable. Improved marketing techniques for native plants have been developed within the horticultural industry.

Grow Native! is a program launched by the Missouri Department of Conservation. This program provides industry-wide branding and tag materials for native plant species. Native landscape brochures, industry and public education programs, and marketing assistance for growers have been developed to support the initiative. Nursery associations and non-governmental conservation organizations also promote native plants and their availability at state or regional levels. The Florida Association of Native Nurseries (FANN) is, for example, a nonprofit organization that promotes native plants. FANN provides publicly available listings for wholesale and retail nurseries and the native plant species that they carry. Other sections of the country provide similar websites. Calflora is a digital library of native and non-native plants found in California, and was instrumental in the formation of the California Native Plant Link Exchange (CNPLX). The CNPLX is a collection of links to websites of regional retail and wholesale nurseries that supply native species.

Sources: Brzuszek and Harkess (2009), Rainer and West (2016)

The Green Future

Native plants and their communities are the true green infrastructure that will sustain us into the future. Creative approaches are needed to utilize existing knowledge and develop new solutions that resonate with people's increasingly urbanized lives. There are opportunities to involve people through citizen science and the love of gardening. We need to ensure that land is managed sustainably for food, timber, and energy production in an era of rapid global change with native species diversity valued and endangered species protected. The role of native plants and seed production in the rural economy needs to be highlighted and supported.

There will need to be a continuum in approaches to maintain native plant diversity across the U.S. based on conservation and careful management of the most important natural areas for plant diversity through restoration of degraded sites to creation of naturalistic plantings in urban areas. In this way it should be possible to guarantee the future of the rich native flora and extraordinary range of landscapes, ecosystems, and vegetation types from desert to rainforest and from Arctic tundra to subtropical swamps. Connectivity is important to enable natural migration and ensure gene flow. The policies, expertise, enthusiasm, and commitment exist to maintain ecologically connected landscapes. Increased collaboration and coordinated action is now required to bring this green future to fruition.

References

Bagley K (2014) Infographic: a field guide to the U.S. environmental movement. Inside climate news, Brooklyn, NY

Balding M, Williams KJH (2016) Plant blindness and the implications for plant conservation. Conserv Biol 30(6):1192–1199

Baughman OW et al (2019) Strong patterns of intraspecific variation and local adaptation in Great Basin plants revealed through a review of 75 years of experiments. Ecol Evol. https://doi.org/10.1002/3c33.5200

BenDor T, Lester TW, Livengood A, Davis A, Yonavjak L (2015) Estimating the size and impact of the ecological restoration economy. PLoS One 10(6):e0128339. https://doi.org/10.1371/journal.pone.0128339. Accessed 30 Oct 2018

Bennett G, Carroll N, Sever K, Neale A, Hartley C (2016) An atlas of ecosystem markets in the United States. Forest Trends, Washington, DC, 19 p

Broadhurst LM, Lowe A, Coates DJ, Cunningham SA, McDonald M, Vesk PA, Yates C (2008) Seed supply for broadscale restoration: maximizing evolutionary potential. Evol Appl 1(4):587–597

Brook BW, Ellis EC, Perring MP, Mackay AW, Blomqvist L (2012) Does the terrestrial biosphere have planetary tipping points? TREE 1664:1–6

Brooklyn Botanic Garden (2019) Brooklyn botanic garden. https://www.bbg.org/. Accessed 11 June 2019

Brzuszek RF, Harkess RL (2009) Green industry survey of native plant marketing in the southeastern United States. HortTechnology 19(1):168–172

Chicago Botanic Garden (CBG) (2019) Budburst. http://wwwbudburstorg/. Accessed 13 June 2019

Diaz S et al (2019) Summary for policymakers of the global assessment report on biodiversity and ecosystem services of the intergovernmental science-policy platform on biodiversity and ecosystem services. IPBES, Paris. 39 p

Ellis EC (2011) Anthropogenic transformation of the terrestrial biosphere. Phil Trans R Soc A 369:1010–1035

Ellis EC, Kaplan JO, Fuller DQ, Vavrus S, Goldewijk KK, Verburg PH (2013) Used planet: a global history. PNAS 110(20):7978–7985

Ellwood ER, Crimmins TM, Miller-Rushing AJ (2017) Citizen science and conservation: recommendations for a rapidly moving field. Biol Conserv 208:1–4

Haefele M, Loomis J, Bilmes LJ (2016) Total economic valuation of the National Park Service lands and programs: results of a survey of the American public. George Wright Forum 33(3):335–345 http://webdoc.agsci.colostate.edu/DARE/PubLinks/NPSTotalEconValue.pdf. Accessed 30 Oct 2018

Havens K (2017) The role of botanic gardens and arboreta in restoring plants: from populations to ecosystems. Chapter 6. In: Blackmore S, Oldfield SF (eds) Plant conservation science and practice. The role of botanic gardens. Cambridge University Press, Cambridge, UK

Havens K, Kramer AT, Guerrant EO (2014) Getting plant conservation right (or not): the case of the United States. Int J Plant Sci 175(1):3–10

Hereford J (2009) A quantitative survey of local adaptation and fitness trade-offs. Am Nat 173(5):579–588

Institute for Applied Ecology (IAE) (2019) Institute for Applied Ecology Sagebrush in Prisons Project. https://appliedeco.org/programs/sagebrush/. Accessed 20 June 2019

Kramer A, Havens K (2009) Plant conservation genetics in a changing world. Trends Plant Sci 14(11):599–607

Kramer AT, Zorn-Arnold B, Havens K (2013) Applying lessons from the US botanical capacity assessment project to achieve 2020 global strategy for plant conservation targets. Ann Mo Bot Gard 99(2):172–179

Leimu R, Fischer M (2008) A meta-analysis of local adaptation in plants. PLoS One 3:e4010

McCormick SJ, Tijian R (2010) A new focus on plant sciences. Science 330:1021

Mooney HA (2010) The ecosystem-service chain and the biological diversity crisis. Philos Trans R Soc B 365(1537):31–39

Obama BH (2015) Remarks by the President on sustainable development goals. The White House, Washington, DC. https://obamawhitehouse.archives.gov/the-press-office/2015/09/27/remarks-president-sustainabledevelopment-goals. Accessed 13 June 2019

Peters CB, Zhan Y, Schwartz MW, Godoy L, Ballard HL (2017) Trusting land to volunteers: how and why land trusts involve volunteers in ecological monitoring. Biol Conserv 208:48–54

Rainer T, West C (2016) Planting in a post-wild world. Designing plant communities for resilient landscapes. Timber Press. Portland, OR, 272 p

Salles JM (2011) Valuing biodiversity and ecosystem services: Why linking economic values with nature? DR 2011-24 Laboratoire Montpelliérain d'Economie Théorique et Appliquée, Montpellier, France

Sgro` CM, Lowe AJ, Hoffmann AA (2011) Building evolutionary resilience for conserving biodiversity under climate change. Evol Appl 4(2):326–337

Shaw S, Schwarz-Ballard J (2017) Cultivating the power of plants to sustain and enrich life: how public gardens can realise our purpose by focusing on the basic human needs universal to diverse audiences. Chapter 8 in: Blackmore S, Oldfield SF (eds) (2017) Plant conservation science and practice. The role of botanic gardens. Cambridge University Press, Cambridge, UK

United Nations (UN) (2019) Sustainable development goals knowledge platform. https://sustainabledevelopment.un.org/?menu=1300. Accessed 13 June 2019

Wandersee J H, Schussler EE 1998. A model of plant blindness. Poster and paper presented at the 3rd Annual Associates Meeting of the 15° Laboratory, Louisiana State University, Baton Rouge, LA

World Economic Forum (2017) The global risks report 2017, 12th edition. http://wef.ch/risks2017. Accessed 17 Mar 2019

Appendix 1: Federal Environmental Laws and Policy

The legal framework for ensuring environmental sustainability and conservation of biodiversity in the U.S. is comprehensive and complicated. The National Environment Policy Act of 1969 provides a broad framework. The majority of land-use regulations relate to general environmental concerns, such as water or air quality. There are many laws and regulations that govern forestry, but there is no overarching forestry policy that ensures sustainability. The following list is not exhaustive but includes the major legislations and policies that relate to native plant species, the habitats where they grow and legal provisions for their conservation and restoration.

Fish and Wildlife Coordination Act of 1934

The Fish and Wildlife Coordination Act (FWCA) is one of the earliest pieces of U.S. federal environmental legislation. Its purpose is to integrate fish and wildlife conservation measures into water resource planning. As amended, it requires consultation with the U.S. Fish and Wildlife Service (FWS) and state wildlife agencies in planning projects that affect streams, lakes, or other water courses, and developing damage mitigation plans.

National Environmental Policy Act 1969

The purposes of the National Environmental Policy Act (NEPA) are to: declare a national policy that will encourage productive and enjoyable harmony between man and his environment, promote efforts that will prevent or eliminate damage to the

S. F. Oldfield et al., *Seeds of Restoration Success*, Springer Earth System Sciences, https://doi.org/10.1007/978-3-319-96974-9

environment and biosphere and stimulate the health and welfare of man, enrich the understanding of the ecological systems and natural resources important to the Nation, and to establish a Council on Environmental Quality (CEQ).

The CEQ operates within the Executive Office of the President to develop and recommend national policies to the President that promote the improvement of environmental quality and meet the Nation's goals.

NEPA requires all federal agencies proposing major actions that may substantially affect the environment to produce an environmental analysis through either environmental assessments or environmental impact statements. The CEQ is required to ensure that federal agencies meet their obligations under NEPA.

Federal Land Policy and Management Act of 1976

The purposes of the Federal Land Policy and Management Act (FLPMA) are to: establish public land policy and guidelines for its administration; provide for the management, protection, development, and enhancement of the public lands; and for other purposes. FLPMA requires that public lands be managed in a manner that will protect the quality of scientific, scenic, historical, ecological, environmental, air and atmospheric, water resource, and archaeological values. This legislation governs the way public lands are administered by the U.S. Department of the Interior, Bureau of Land Management (BLM).

National Forest Management Act of 1976

The National Forest Management Act (NFMA) governs the administration of the National Forest System and requires the USFS to create a land and resources management plan for each national forest. The agency must consider each unit under the principles of multiple use and sustained yield, taking into account timber, watershed management, recreation, and other forest uses. Public participation is required for the creation of such management plans.

NFMA relates to wider forest management, stating that "*inasmuch as the majority of the Nation's forests and rangeland is under private, state, and local governmental management and the Nation's major capacity to produce goods and services is based on these non federally managed renewable resources, the federal Government should be a catalyst to encourage and assist these owners in the efficient long-term use and improvement of these lands and their renewable resources consistent with the principles of sustained yield and multiple use.*"

The Rural Development Act of 1972

The purpose of the Rural Development Act is to revitalize and develop rural areas and to help foster a balance between rural and urban America. This Act named the United States Department of Agriculture (USDA) as the lead federal coordinator for rural policy.

The Soil and Water Resources Conservation Act of 1977

The Soil and Water Resources Conservation Act (RCA) as amended provides the USDA with broad strategic assessment and planning authority for the conservation, protection, and enhancement of soil, water, and related natural resources. Through this, the USDA appraises the status and trends of soil, water, and related resources on non-federal land and assesses their capability to meet present and future demands; evaluates current and needed programs, policies, and authorities; and develops a national soil and water conservation program to give direction to USDA soil and water conservation activities.

The National Historic Preservation Act of 1966

This Act recognizes the need for preservation of irreplaceable historical heritage in the public interest so that its vital legacy of cultural, educational, aesthetic, inspirational, economic, and energy benefits will be maintained and enriched for future generations of Americans. It provides the means to do so, including the maintenance of a National Register of Historic Places composed of districts, sites, buildings, structures, and objects significant in American history, architecture, archaeology, engineering, and culture. It also designates "National Historic Landmarks" to be included on the National Register.

Endangered Species Act of 1973

The Endangered Species Act (ESA) provides a program for the conservation of threatened and endangered plants and animals and the habitats in which they are found. The U.S. Fish and Wildlife Service (FWS) of the Department of the Interior maintains the list of endangered species and threatened species. The law requires federal agencies, in consultation with the FWS and/or the Department of Commerce, National Oceanic and Atmospheric Administration (NOAA), to ensure that actions they authorize, fund, or carry out are not likely to jeopardize the continued existence of any listed species or result in the destruction or adverse modification of designated critical habitat of such species. The ESA also prohibits any action that causes a "taking" of any listed species of endangered fish or wildlife. Likewise, import, export, interstate, and foreign commerce of listed species are all generally prohibited.

The Lacey Act of 1900

The Lacey Act was the first federal law protecting wildlife in the U.S. The Act as amended prohibits trade in wildlife, fish, and plants that have been illegally taken, possessed, transported, or sold. It enforces civil and criminal penalties for the illegal trade in species, underscoring federal, state, and foreign laws. The Act regulates the import of protected species and prevents the spread of invasive, non-native species.

Clean Air Act of 1963

The Clean Air Act regulates air emissions. First passed in 1963, amendments in 1970 provided the basis for federal controls. This Act authorizes the U.S. Environmental Protection Agency (EPA) to establish National Ambient Air Quality Standards (NAAQS) to protect public health and the environment. The 1990 amendments to the Clean Air Act were intended to meet unaddressed or insufficiently addressed problems such as acid rain, ground-level ozone, stratospheric ozone depletion, and air toxics.

Clean Water Act of 1972

The basis of the Clean Water Act was enacted in 1948 and was called the Federal Water Pollution Control Act. This Act was significantly reorganized and expanded in 1972. The Act establishes the basic structure for regulating discharges of pollutants into the waters of the U.S. It gave EPA the authority to implement pollution control programs such as setting wastewater standards for industry.

The Comprehensive Environmental Response, Compensation, and Liability Act of 1980

The Comprehensive Environmental Response, Compensation, and Liability Act commonly known as CERCLA or Superfund provides broad federal authority to respond directly to releases or threatened releases of hazardous substances that may endanger public health or the environment. CERCLA was amended by the Superfund

Amendments and Reauthorization Act (SARA) on October 17, 1986 and the Brownfields Utilization, Investment, and Local Development Act (BUILD) of March 23, 2018.

Energy Policy and Conservation Act of 1975

This Energy Policy and Conservation Act (ECPA) provided a comprehensive approach to federal energy policy, following the 1973 oil crisis. The primary goals of EPCA were to increase energy production and supply, reduce energy demand, provide energy efficiency, and provide the executive branch with additional powers to respond to disruptions in energy supply. The EPCA was amended and enacted October 23, 2018.

Global Change Research Act of 1990

The Global Change Research Act (GCRA) requires research into global warming and related issues. It requires a report to Congress every 4 years on the environmental, economic, health, and safety consequences of climate change. Furthermore, the GCRA mandates that the research findings of the Committee and off ederal agencies and departments be made available to the EPA and all federal agencies and departments.

Energy Independence and Security Act of 2007

This Act aims to move the U.S. toward greater energy independence and security. It introduces measures to expand the production of renewable fuels, reduce U.S. dependence on oil, increase energy security, and address climate change. In developing restoration activities, management strategies and adaptation strategies, the use of native plant species for sequestering covered greenhouse gas in each ecosystem is emphasized.

President's Climate Action Plan, Executive Office of the President, June 2013

This broad-based plan was developed to cut the carbon pollution that causes climate change and affects public health. It included measures by the Obama Administration to help state and local governments strengthen roads, bridges, and shorelines to better protect homes, businesses, and ways of life from severe weather events.

This policy is rescinded by the Executive Order 3349: American Energy Independence, March 28, 2017.

National Fish, Wildlife & Plants Climate Adaptation Strategy (NFWPCAS 2012)

This Strategy was developed by FWS, along with NOAA and other federal agencies, state agencies, and tribal partners. The Strategy identifies seven goals to help fish, wildlife, plants, and ecosystems cope with the impacts of climate change.

National Strategy to Promote The Health Of Honey Bees And Other Pollinators (Pollinator Health Task Force, 2015)

This Strategy expands and adds to actions already being undertaken by federal departments and agencies to reverse pollinator losses and restore populations to healthy levels. It focuses on both immediate and long-term changes that can be made to improve the well-being of pollinator populations. Consequently, the Strategy addresses the many factors impacting pollinator health, including land-use practices, declining forage and nesting resources, pests and diseases, pesticides, and bee biology.

Executive Order 13751 of December 5, 2016: Safeguarding the Nation from the Impacts of Invasive Species

This Order states that it is the policy of the United States to prevent the introduction, establishment, and spread of invasive species, as well as to eradicate and control populations of invasive species that are established. It amends Executive Order 13112 and directs actions to continue coordinated federal prevention and control efforts related to invasive species. This Order maintains the National Invasive Species Council (Council) and the Invasive Species Advisory Committee, expands the membership of the Council, and strengthens coordinated federal action.

Executive Order 13423 of January 24, 2007: Strengthening Federal Environmental, Energy, and Transportation Management

This Order promotes and sets targets for federal agencies on issues such as energy efficiency and reduction of greenhouse gas emissions, water consumption, and use of sustainably produced and recycled products.

Executive Order 13514 of October 5, 2009: Federal Leadership in Environmental, Energy, and Economic Performance

Building on Executive Order 13423, this Order calls for an inventory of all federal greenhouse gas emissions and expands the definition beyond direct greenhouse gas emissions from sources that are owned or controlled by the federal agencies. Of particular significance to acquisition activities, Executive Order 13514 includes an increased emphasis on high performance sustainable buildings including planning and targets for zero net energy buildings.

This Order is revoked by Executive Order 3349: American Energy Independence, March 28, 2017.

Executive Order 13508 of May 12, 2009: Chesapeake Bay Protection and Restoration

This Order recognizes the Chesapeake Bay as a national treasure and calls on the federal government to lead a renewed effort to restore and protect the nation's largest estuary and its watershed. Priority strategic actions were defined as restoring clean water, recovering habitats, sustaining fish and wildlife, conserving land, and increasing public access.

Executive Order 13547 of July 19, 2010: Stewardship of the Ocean, Our Coasts, and the Great Lakes

This Order establishes a national policy to ensure the protection, maintenance, and restoration of the health of ocean, coastal, and Great Lakes ecosystems and resources; enhance the sustainability of ocean and coastal economies; preserve maritime heritage; support sustainable uses and access; provide for adaptive management to enhance our understanding of and capacity to respond to climate change and ocean acidification; and coordinate with national security and foreign policy interests. This Order also provides for the development of coastal and marine spatial plans that build upon and improve existing federal, state, tribal, local, and regional decision making and planning processes. This Order was revoked on June 19, 2018 by the Executive Order Regarding the Ocean Policy to Advance the Economic, Security, and Environmental Interest of the United States.

Executive Order 13653 of November 1, 2013: Preparing the United States for the Impacts of Climate Change

This Order recognized that impacts of climate change, including extreme weather events are already affecting communities, natural resources, ecosystems, economies, and public health across the U.S. Managing these risks requires deliberate preparation, close cooperation, and coordinated planning by the Federal Government, as well as by stakeholders, to improve climate preparedness and resilience. Building on earlier Executive Orders and programs including the U.S. Global Change Research Program (USGCRP), established under the Global Change Research Act of 1990 the Federal Government will continue to support scientific research, observational capabilities, and assessments relating to climate change.

Agencies should promote: (1) engaged and strong partnerships and information sharing at all levels of government; (2) risk-informed decision making and the tools to facilitate it; (3) adaptive learning, in which experiences serve as opportunities to inform and adjust future actions; and (4) preparedness planning.

This policy is rescinded by Executive Order 3349: American Energy Independence, March 28, 2017.

Interior Department Secretarial Order 3330 Improving Mitigation Policies and Practices of the Department of the Interior, 2013

This Order promoted mitigation policies and practices that (a) more effectively avoid, minimize, and compensate for the impact of development on Department-managed lands and resources; (b) provide better information and greater predictability to project proponents and land managers; (c) improve the resilience of our Nation's resources in the face of climate change; (d) encourage more strategic conservation investments in lands and other resources; and (e) increase compensatory mitigation effectiveness, durability, transparency, and consistency.

This Order was revoked following Executive Order 3349: American Energy Independence, March 28, 2017.

Interior Department Secretarial Order 3336 Rangeland Fire Prevention, Management & Restoration, 2015

This order sets out enhanced policies and strategies for preventing and suppressing rangeland fire and for restoring sagebrush landscapes impacted by fire across the West. These actions are essential for conserving habitat for the greater sage-grouse and other wildlife species and economic activity, such as ranching and recreation, associated with the sagebrush-steppe ecosystem in the Great Basin region.

Executive Order 3349: American Energy Independence March 28, 2017

This order directs the Department to review all existing regulations, orders, guidance documents, policies, and any other similar actions (including those referred to above) that potentially burden the development or utilization of domestically produced energy resources. The objective of the review is to identify agency actions that unnecessarily burden the development or utilization of the Nation's energy resources and support action to appropriately and lawfully suspend, revise, or rescind such agency actions as soon as practicable.

Appendix 2: Internationally Important Wetland Sites of the U.S. Listed on the Ramsar Convention

State	Ramsar site	Habitats
Alaska	Izembek Lagoon National Wildlife Refuge	Lagoons and marshes; one of the world's largest eelgrass beds
Arkansas	Cache-Lower White Rivers	Lower Mississippi Alluvial Valley Ecosystem with over 70 tree species
California	Tijuana River National Estuarine Research Reserve	Estuary and coastal lagoon with sand dunes, beaches, vernal pools, tidal channels, mudflats, and coastal sage scrub
California	Tomales Bay	Marine-coastal wetland with geomorphologically dynamic estuaries, eelgrass beds (*Zostera marina*), sand dune systems, and restored emergent tidal marshes
California	Bolinas Lagoon	Tidal embayment of open water, mudflats, and marsh
California	Laguna de Santa Rosa Wetland Complex	Creeks, ponds, marshes, vernal pools, swales, floodplains, riparian forest and grassland. Endemic plants such as Sonoma sunshine (*Blennosperma bakeri*)
California	San Francisco Bay/ Estuary (SFBE)	Largest estuary on the Pacific Coast of the U.S.
California	Grassland Ecological Area (GEA)	Semipermanent and permanent marshes, riparian corridors, vernal pool complexes, wet meadows, native uplands, and grasslands featuring alkali sacaton (*Sporobolus airoides*) and the endemic delta button celery (*Eryngium racemosum*)
Connecticut	Connecticut River Estuary and Tidal River Wetlands Complex	Open water; fresh, salt, and brackish tidal wetlands; floodplains, river islands, beaches, and dunes
Delaware, New Jersey	Delaware Bay Estuary	More than 70 separate protected wetland areas of varying salinity supporting various marsh communities

(continued)

S. F. Oldfield et al., *Seeds of Restoration Success*, Springer Earth System Sciences, https://doi.org/10.1007/978-3-319-96974-9

(continued)

State	Ramsar site	Habitats
Florida	Everglades National Park	Freshwater and wet prairies, islands of tropical hardwood trees, saltmarshes, mangrove forests, beach and dune complexes, and brackish water estuaries with over 1,000 vascular plants
Florida	Corkscrew Swamp Sanctuary	Pine flatwoods, wet prairies, cypress swamps, and marshes with 600 year-old bald cypress (*Taxodium distichum*) comprise the largest remaining virgin bald cypress forest in North America
Florida	Pelican Island National Wildlife Refuge	Brackish water estuarine system
Florida, Georgia	Okefenokee National Wildlife Refuge	Swamp forest with old-growth cypress and slash pine (*Pinus elliottii*)
Hawaii	Kawainui and Hamakua Marsh Complex	Ancient freshwater fishpond located in the caldera of the Koolau shield volcano
Illinois	The Emiquon Complex	Bottomland lakes, side channels, sloughs, marshes, bottomland hardwood forests, and wet, mesic, and dry prairies
Illinois	Cache River and Cypress Creek Wetlands	Contains rare examples of forested swamps dominated by bald cypress and water tupelo (*Nyssa aquatica*)
Illinois	Sue and Wes Dixon Waterfowl Refuge at Hennepin & Hopper Lakes	Backwater lakes, marshes, wet prairie, savanna, and forest that has been restored to a landscape closely resembling presettlement conditions
Illinois, Wisconsin	Chiwaukee Illinois Beach Lake Plain	Coastal dune and swale ecosystem with eastern prairie fringed orchid (*Platanthera leucophaea*)
Kansas	Quivira National Wildlife Refuge	Inland salt marsh with mixed grass prairie
Kansas	Cheyenne Bottoms	A natural elliptical basin on the Great Plains with shallow marshes
Louisiana	Catahoula Lake	Large, poorly drained depression subject to artificially controlled, seasonal fluctuations
Michigan	Humbug Marsh	The last stretch of natural shoreline on the U.S. mainland connecting the lower and upper Great Lakes with freshwater lagoons and seasonal marshes
Minnesota, Iowa, Wisconsin, Illinois	Upper Mississippi River Floodplain Wetlands	Main and side channel habitats, large shallow to moderately deep backwater marshes, floodplain forests, and shrub-dominated communities
Nevada	Ash Meadows National Wildlife Refuge	A unique oasis ecosystem of streams, pools, and wet meadows, formed around springs supplying a tributary of the Amargosa River with nine endemic plant species
New Jersey	Edwin B. Forsythe National Wildlife Refuge	Coastal salt meadows, flooded woodlands, and open bays and channels

(continued)

(continued)

State	Ramsar site	Habitats
New Mexico	Roswell Artesian Wetlands	Springs, seeps, and sinkhole lakes located along the Pecos River but largely fed by groundwater discharged from the Roswell Basin
Ohio	Wilma H. Schiermeier Olentangy River Wetland Research Park	Freshwater, tree-dominated freshwater wetlands, permanent rivers/streams/creeks, seasonal/intermittent freshwater marshes, and permanent freshwater marshes
Pacific Region	Palmyra Atoll National Wildlife Refuge	Coral reefs, permanent shallow marine waters, and intertidal forested wetlands
South Carolina	Francis Beidler Forest	Swamp forest with the largest remaining virgin stand of bald cypress (*Taxodium distichum*) and tupelo gum (*Nyssa sylvatica*) trees in the world. 300 plant species including rare orchids
South Carolina	Congaree National Park	Freshwater swamp forests, seasonal sloughs, forested peatlands, permanent and seasonal creeks, freshwater lakes, and shrub-dominated wetlands, containing the largest remaining example of old-growth bottomland hardwood forest in North America
South Dakota	Sand Lake National Wildlife Refuge	Freshwater cattail marsh
Texas, Louisiana	Caddo Lake	Shallow, heavily vegetated lake waters with sloughs, swamps, backwaters and hardwood forests with bald cypress
Vermont	Missisquoi Delta and Bay Wetlands	Contains the largest contiguous floodplain forest in Vermont and unique habitat types such as the Maquam Bog.
Virginia	Chesapeake Bay Estuarine Complex	Sand beaches, dunes, mudflats, open water with submerged beds of aquatic vegetation, intertidal marshes, freshwater marshes, and lakes
Wisconsin	Door Peninsula Coastal Wetlands	Includes interdunal wetlands and northern wet-mesic forest. Rare dwarf lake iris (*Iris lacustris*)
Wisconsin	Horicon Marsh	One of the largest cattail (*Typha*) marshes in the world
Wisconsin	Kakagon and Bad River Sloughs	Sloughs, bogs, and coastal lagoons with the largest natural wild rice (*Zizania palustris*) bed on the Great Lakes.

References

The Ramsar Convention Secretariat (2014) Sites and countries. https://www.ramsar.org/sites-countries. Assessed 2 June 2019.

Index

A
Agricultural land management, 64
Agricultural Research Service's Western Regional Plant Introduction Station (ARS WRPIS), 81
Agriculture
  coastal plains, 67
  cropland, 66
  crop production, 69
  drought, 69
  ecological changes, 67
  economic and social consequences, 69
  grain farming, 68
  grasses, 67
  Great Plains, 68
  intercultivation methods, 68
  land tenure, 66
  livestock management, 69
  meadow plants, 67
  open-ranching economy, 68
  soil conservation, 69
Aichi Targets, 3
"American Beauties" program, 14
The American Public Gardens Association (APGA), 55
Arrowleaf balsamroot, 92

B
Basalt milkvetch, 92
Bioenergy, 15, 16
Botanic gardens, 110, 111
  and arboreta, in the U.S., 52
  BGCI, 55
  Chicago Botanic Garden, 53
  in situ and *ex situ* plant conservation, 51
  Missouri Botanical Garden, 53
  NCGRP, 55
  NTBG, 54
  PlantSearch database, 51
  Plants of Concern Program, 49
  Rancho Santa Ana Botanic Garden, 54
  as seed banks, 50
Botanic Gardens Conservation International (BGCI), 51, 55
Budburst, 111
Bureau of Land Management (BLM), 61, 78

C
Cape May Plant Materials Center (CMPMC), 95
Carbon storage, 25, 37, 63, 64
Cellulose, 15
Center for Plant Conservation (CPC), 51
Citizen science, 111–113
Clean Air Act, 118
Clean Water Act, 118
Climate change, 2
  drought, 19
  temperature and rainfall, 19
  U.S. forests, 19
  wildfire risk, 19
Climate Change Adaptation Plans, 3
Coastal wetlands, 94–96
Collaborative Forest Landscape Restoration (CFLR) sites, 78
Colorado Plateau Native Plant Program (CPNPP), 93
Comprehensive Environmental Response, Compensation, and Liability Act (CERCLA), 118

S. F. Oldfield et al., *Seeds of Restoration Success*, Springer Earth System Sciences, https://doi.org/10.1007/978-3-319-96974-9

Conservation, 73–74
Conservation assessment, 42
  awareness and behavior, 44
  conservation status rank, 42, 43
  ex situ (*see Ex situ* conservation)
  in situ (*see In situ* conservation)
Conservation genetics, 49
Conservation Reserve Program (CRP), 64
Crested Butte Wildflower Festival, 8, 9
Cropland, 65
Crop wild relatives (CWR), 11

D
Dawes Act, 68
Desert restoration, 90–94
Deserts
  Chihuahuan Desert, 36
  cold, 34
  Mojave Desert, 35
  restoration, 90–94
  Sagebrush, 34
  salt, 34
  Sonoran Desert, 35
  species, sagebrush, 34
  vegetation, 36
Dust Bowl, 69
Dusty maiden, 92

E
Eastern Temperate Forests Ecoregion, 27
Ecological resilience, 3, 4
Ecological restoration, 77–81, 84, 86, 87, 93, 95, 96, 98, 104–106
Ecoregions
  deserts (*see* Deserts)
  forests (*see* Forests)
  grasslands, 33
  Level I ecoregions, in US, 25–27
  Level II ecoregions, 27
  Level III ecoregions, 27
  North American ecoregion classification system, 25
  vegetation types, in US, 28
  wetlands, 36, 37
  woodlands, 32, 33
Ecosystems
  and biodiversity, 3
  carbon sequestration, 3
  and carbon storage, 37
  changing ecosystems, 38
  ecoregions (*see* Ecoregions)
  environmental change, 38
Ecosystem services
  aesthetic and passive use, 14
  carbon dioxide and green plants, 9
  economic value, 16
  insect pollination, 10
  native plant diversity, 9
  native plants, 9
  native vegetation, 11
  and provisioning, 10
  watershed vegetation, 9
Endangered Species Act (ESA), 43, 45, 50, 53, 117
Energy Policy and Conservation Act (EPCA), 119
Executive Order 13653, 4
Ex situ conservation
  APGA, 55
  botanic gardens and arboreta, 50, 52
  Chicago Botanic Garden, 53
  concentrated conservation efforts, 54
  genetic factors, 55
  high quality *ex situ* collections, 50
  Missouri Botanical Garden, 53
  NCGRP, 55
  NTBG, 54
  plant rich habitats, 50
  PlantSearch database, 51
  Rancho Santa Ana Botanic, 54
  Texas snowbells, 51
Extinct plants, 42
Extreme weather events, 2–4

F
Federal land management, 61–64
Federal Land Policy and Management Act (FLPMA), 61, 116
Fish and Wildlife Coordination Act (FWCA), 115
Fish and Wildlife Service (FWS), 61, 95
Forestland, 65
Forest restoration program, 78, 88
Forestry, 71–73
Forests
  Appalachian forests, 30
  Eastern deciduous forests, 29
  Eastern Temperate Forests Ecoregion, 29
  forested land, in US, 27
  Marine West Coast Forest Ecoregion, 31
  Northern Forests Ecoregion, 27, 29
  Northwestern Forested Mountains Ecoregion, 30
  restoration, 87–89
  tropical humid forest ecoregion, 30

and woodland, 31
Forest Service (USFS), 61
Forest Service's Forest Products Laboratory, 73
Forward-thinking management, 106

G
Gap Analysis Program (GAP), 47
General Allotment Act, 68
Gooseberryleaf globemallow, 92
Governor's Office of Storm Recovery (GOSR), 95
Grasslands, 33
Great Basin Sagebrush Project, 108
Green infrastucture, 25

H
Habitat loss, 17
Hawaii Plant Extinction Prevention (PEP) Program, 54
Housing and Urban Development (HUD), 95

I
In situ conservation
  California, 48
  challenge, 50
  Colorado Natural Heritage Program, 48
  ecological restoration, 50
  ESA, 45
  major protected areas, 45, 46
  in natural habitats, 45
  Redwood National Park, 48
  Research Natural Areas (RNAs), 48
  state and local governments, 47
  threatened plants, 44
Invasive species, 16–18, 21, 22

L
Lacey Act, 118
Land degradation, 70, 71
Land ownership, 59–60
Landscapes, 1
Land trusts, 66
Limestone hawksbeard, 92

M
Medicinal plants
  American ginseng, 13, 11
  black cohosh/snakeroot, 13
  Cherokee medicine, 12
  Echinacea cultivation, 13
  golden seal, use of, 12
  healthcare and pharmaceutical developments, 11
  herbs, 12
  purple cornflower, 13
Mid-Atlantic Regional Seed Bank (MARSB), 95
Mitigation, 41, 78, 80, 94
Mojave desert tortoise, 94

N
Nanocrystalline cellulose (NCC), 15
National Center for Genetic Resources Preservation (NCGRP), 55, 82
National Conservations Lands, 61
National Environmental Policy Act (NEPA), 78, 115
National Forest Management Act (NFMA), 72, 116
National Parks, 61, 73
National Park Service (NPS), 61, 93
National Plant Germplasm System (NPGS), 81
National seed strategy, 97–99, 107–109
The National Tropical Botanical Garden (NTBG), 54
National Wetlands Inventory (NWI), 36
National Wildlife Refuge System, 61
Native plant
  climate change, 19
  conservation (*see* Conservation assessment)
  diversity, 41
  diversity hotspots (*see* Plant biodiversity hotspots)
  for economy, 15, 16
  for ecosystem services, 9, 10
  for food, 10
  habitat loss, 17
  invasive species, 17–18
  for medicine (*see* Medicinal plants)
  native seed, 7
  for ornamental and landscape use, 13–14
  overexploitation, 18
  resource, the U.S., 8
  scientific documentation, native U.S. flora, 8
  under threat, 16–17
Native plant communities, 1, 3, 4
Native plant production, 80
Natural Resources Conservation Service (NRCS), 69, 83

NatureServe Network, 43
New England Wild Flower Society (NEWFS), 95
Nineleaf biscuitroot, 92
Non-federal land management, 64–66
North Carolina Botanic Garden (NCBG), 95
Northwest Forest Plan (NWFP), 72

O
Oak savannas, 90
Organic Administration Act of 1897, 63
Ornamental horticulture, 14
Overexploitation, plants, 18

P
Pastureland, 65
Perennial bioenergy plants, 16
Plant biodiversity hotspots
  California Floristic Province, 20
  California hotspot, 20
  Edwards Plateau, 21
  endemic tree species, 20
  Hawaii, 21
  human population, 21
  national and global botanical, 22
  North American Coastal Plain, 21
Plant blindness, 103, 110, 111
Plant Conservation Alliance (PCA), 56
Plant diversity, 103, 107, 113
Plant Materials Centers (PMCs), 83
Pollinators, 96–97
Prairie restoration, 89
Prairie vegetation, 89
Provisioning services
  CWR, 11
  economic contribution, cultivated sunflower, 10
  sunflower, 10
  native vegetation, 11
  plant species, 10
Purchase-of-development rights (PDRs), 66

R
Ramsar Convention, 125–127
Rangeland, 65
Recalcitrant species, 54
Recreation, 73–74
Resilience, 4
Restoration, 59
  agricultural seed production, 82–84
  BLM, 78
  botanic gardens and ecological, 79
  coastal wetlands, 94–96
  collection and storage, seed, 81–82
  deserts, 90–94
  ecological resilience in New York, 95, 96
  ecoregions, 85, 87
  federal agencies, 78
  food and timber production, 80
  forests, 87–89
  Great Basin, 92–94
  land management, 77
  national policies, 81
  National Seed Strategy, 97–99
  Native Plant Materials Development Program, 80
  oak savannas, 90
  plant materials, 78
  plant species, 84
  pollinators, 96–97
  prairies, 89–90
  as seed zones, 86–87
  SER, 77
  social, regulatory and ecological condition, 77
Royal penstemon, 92

S
Sagebrush, 34, 35
Sage grouse, 35
Saguaro, 22
Seed banks, 49–51, 53–56
Seed collection, 79, 81, 86, 95
Seeds
  banking facilities, 4
  description, 2
Seeds of Success (SOS), 51, 81
As Seed zones, 86–87
Shaggy fleabane, 93
Silky lupine, 93
Soil and Water Resources Conservation Act (RCA), 117
Soil Conservation Service (SCS), 69
Sulphur-flower buckwheat, 93
Sustainable Development Goals (SDGs), 104

T
Texas wild rice, 11
Threatened plant species, 16–17, 21

U
UN Biodiversity Targets, 3
UN Sustainable Development Goals (SDGs), 104

Urban landscaping, 112
Urbanization, 75
US Geological Service (USGS), 93
US Global Change Research Program (USGCRP), 122

V
Vegetation types, 1

W
Watershed vegetation, 9
Wetlands
  carbon storage, 37
  defined, 36
  Everglades, 36, 37
  NWI, 36
  salt marsh, 36
  terrestrial and aquatic ecosystems, 37
Wetland sites, US, 125–127
Whitebark pine, 18
Wildlife Habitat Incentives Program (WHIP), 90
Wild plant species, 7
Woodlands
  chaparral/sclerophyllous shrubland, 32
  and forest, 31
  oak woodland, 32, 33
  pinyon-juniper woodland extends, 32
  types, 32

Scientific Index

A

Abies, 19
Abies amabilis, 31
Abies balsamea, 27
Abies balsamea var. *phanerolepis,* 19
Abies fraseri, 19
Abies grandis, 31
Abies lasiocarpa, 31
Acer, 72
Acer grandidentatum, 21
Acer rubrum, 29
Acer saccharum, 29
Achnatherum hymenoides, 81
Adelges piceae, 19
Adelges tsugae, 19
Adenostoma fasciculatum, 32
Agave, 16
Agrilus planipennis, 29
Agropyron cristatum, 65
Ambrosia dumosa, 35
Amorpha canescens, 97
Anthoxanthum aristatum, 67
Apis mellifera, 10
Aquilegia coerulea, 9
Arbutus xalapensis, 21
Arctostaphylos spp., 32
Artemisia, 34, 81, 91
Artemisia californica, 32
Artemisia tridentata, 34
Artemisia tridentata ssp. *parishii,* 34
Artemisia tridentata ssp. *tridentata,* 34, 91
Artemisia tridentata ssp. *vaseyana,* 34
Artemisia tridentata ssp. *wyomingensis,* 34, 91
Astragalus filipes, 92, 97
Atriplex-Sarcobatus spp., 93

B

Balsamorhiza sagittata, 92
Baptisia, 14
Betula papyrifera, 29
Bison bison, 74
Blennosperma bakeri, 125
Bouteloua gracilis, 94
Bromus hordaceus, 67
Bromus rubens, 94
Bromus tectorum, 34, 67, 90
Bursera simaruba, 30

C

Carnegiea gigantea, 22
Carya spp., 21
Castanea dentata, 29, 88
Castilleja spp., 8
Caulophyllum thalictroides, 12
Ceanothus cuneatus, 32
Celtis spp., 21
Centaurea solstitialis, 17
Centrocercus urophasianus, 3, 35, 91
Cercis canadensis, 14
Cercocarpus traskiae, 54
Chaenactis douglasii, 92
Chrysobalanus icaco, 30
Cimicifuga racemosa, 12
Cirsium pitcheri, 53
Coleogyne ramosissima, 93
Coreopsis spp., 21
Cornus spp., 30
Corylus spp., 10
Crepis intermedia, 92
Cronartium ribicola, 18

S. F. Oldfield et al., *Seeds of Restoration Success*, Springer Earth System Sciences, https://doi.org/10.1007/978-3-319-96974-9

Croton alabamensis var. texensis, 21
Croton obovatus, 21
Cucurbita spp., 10
Cupressus spp., 32

D
Dalea purpurea, 97
Denroctonus ponderosae, 18

E
Echinacea, 14
Echinacea angustifolia, 13
Echinacea paradoxa, 13
Echinacea purpurea, 13
Elymus elymoides, 81
Endothia parasitica, 29
Ephedra spp., 94
Eremothera boothii, 94
Erigeron pumilus, 93
Eriogonum umbellatum, 93
Eryngium racemosum, 125
Erythranthe pericaulis, 42

F
Fagus, 72
Fagus grandifolia, 29
Festuca, 33
Flourensia cernua, 36
Franklinia alatamaha, 42
Fraxinus americana, 29

G
Gaillardia spp., 21
Gaultheria shallon, 48
Gentiana spp., 8
Gopherus agassizii, 94

H
Hamamelis spp., 21
Helianthus annuus, 10
Helianthus paradoxus, 10, 51
Helianthus quinquenervis, 8
Helianthus spp., 10
Heliomeris multiflora, 94
Hesperostipa comata, 81
Holcus lanatus, 67
Hydrangea quercifolia, 14
Hydrastis canadensis, 12

I
Ipomopsis aggregata, 94
Iris lacustris, 127

J
Juglans spp., 11
Juniperus ashei, 21
Juniperus coahuilensis, 32
Juniperus deppeana, 32
Juniperus monosperma, 32
Juniperus osteosperma, 32
Juniperus scopulorum, 32
Juniperus virginiana, 79

L
Larix laricina, 27
Larrea tridentata, 35, 94
Leymus cinereus, 81
Liriodendron tulipifera, 14
Lolium perenne, 67
Lomatium triternatum, 92
Lupinus sericeus, 93
Lupinus spp., 8

M
Machaeranthera canescens, 92
Malacothrix glabrata, 94
Mespilus canescens, 53
Micrathene whitneyi, 22
Mirabilis laevis, 94

N
Nyssa aquatica, 126
Nyssa sylvatica, 127

O
Olneya tesota, 36
Orbexilum macrophyllum, 42
Osmia spp., 97

P
Panax quinquefolius, 11
Panicum virgatum, 16
Parkinsonia spp., 35
Parthenocissus quinquefolia, 14
Pennisetum ciliare, 22
Penstemon speciosus, 92

Penstemon spp., 8
Phaseolus spp., 10
Picea engelmannii, 31
Picea glauca, 27
Picea mariana, 27
Picea rubens, 29
Picea sitchensis, 31
Pinus albicaulis, 18
Pinus attenuata, 32
Pinus banksiana, 27
Pinus cembroides, 32
Pinus contorta, 31
Pinus edulis, 32
Pinus elliottii, 126
Pinus monophylla, 32
Pinus palustris, 30, 88
Pinus ponderosa, 31, 70
Pinus resinosa, 29
Pinus spp., 10
Pinus strobus, 29
Plantago ovata, 94
Platanthera leucophaea, 126
Platanus spp., 21
Poa pratensis, 67
Poa secunda, 81
Polystichum, 48
Populus, 72
Populus balsamifera, 29
Populus spp., 16, 87
Populus tremuloides, 29
Potentilla fruticosa, 14
Prosopis glandulosa, 21
Prosopis spp., 35
Prunus spp., 11
Pseudotsuga macrocarpa, 32
Pseudotsuga menziesii var. *glauca,* 31
Pseudotsuga menziesii var.*menziesii,* 31

Q
Quercus spp., 10, 29, 90
Quercus virginiana, 30

R
Rhododendron macrophyllum, 48
Rudbeckia laciniata, 14

S
Salix spp., 16, 87
Sanguinaria canadensis, 12
Sequoiadendron giganteum, 20
Sequoiadendron gigantum, 74
Sequoia sempervirens, 20, 31
Solidago, 96
Sphaeralcea ambigua, 94
Sphaeralcea grossulariifolia, 92
Sphaeralcea parvifolia, 93
Sporobolus airoides, 125
Streptanthus bracteatus, 21
Strix occidentalis caurina, 72
Styrax platanifolius ssp. *texanus,* 51
Styrax texana, 21, 51
Swietenia mahogoni, 30
Symphyotrichum, 96

T
Taxodium distichum, 126, 127
Taxodium spp., 21
Taxus, 54
Thismia americana, 42
Thuja plicata, 31
Tilia spp., 29
Torreya, 54
Trifolium repens, 67
Triticum aestivum, 68
Tsuga, 72
Tsuga caroliniana, 19
Tsuga heterophylla, 31
Typha, 127

V
Vaccinum spp., 10

Z
Zea mays, 10
Zenaida asiatica, 22
Zizania aquatica, 11
Zizania palustris, 11, 127
Zizania spp., 10
Zizania texana, 11, 51
Zostera marina, 125

The manufacturer's authorised representative in the EU is Springer Nature Customer Service Centre GmbH, Europaplatz 3, 69115 Heidelberg, Germany. If you have any concerns regarding our products, please contact ProductSafety@springernature.com

Printed and bound by CPI Group (UK) Ltd, Croydon, CR0 4YY
30/09/2026
02225610-0003